Colossal Book of WORD PLAY

Martin Gardner
with Ken Jennings

PUZZLE WRIGHT PRESS

An imprint of Sterling
Publishing Co., Inc.

www.puzzlewright.com

To Ross Eckler, for his books on advanced wordplay
and for his distinguished career as the editor of *Word Ways*,
the world's leading journal of linguistic recreations.

Vista logo courtesy of UNA-USA.

Puzzlewright Press and the distinctive Puzzlewright Press logo
are registered trademarks of Sterling Publishing Co., Inc.

2   4   6   8   10   9   7   5   3   1

Published by Sterling Publishing Co., Inc.
387 Park Avenue South, New York, NY 10016
© 2010 by Martin Gardner
Distributed in Canada by Sterling Publishing
c/o Canadian Manda Group, 165 Dufferin Street
Toronto, Ontario, Canada M6K 3H6
Distributed in the United Kingdom by GMC Distribution Services
Castle Place, 166 High Street, Lewes, East Sussex, England BN7 1XU
Distributed in Australia by Capricorn Link (Australia) Pty. Ltd.
P.O. Box 704, Windsor, NSW 2756, Australia

Manufactured in the United States of America
All rights reserved

Sterling ISBN 978-1-4027-6503-2

For information about custom editions, special sales, premium and
corporate purchases, please contact Sterling Special Sales Department
at 800-805-5489 or specialsales@sterlingpublishing.com.

CONTENTS

FOREWORD
by Ken Jennings

I t's been said that the early records by the legendary 1960s art-rock pioneers the Velvet Underground sold only a few thousand copies, but that every single person who bought one started a band. If this is true, then Martin Gardner is the Velvet Underground of 20th-century thinkers: he may never have been a household name, but his prolific writing inspired a generation or two of mathematicians and scientists. His monthly "Mathematical Games" feature ran in *Scientific American* magazine for 25 years, creating a sort of proto-Internet in its devoted community of readers, who would dive headlong into every recreational math puzzle Gardner proposed, exchanging notes and discoveries on fractals or public-key cryptography or the lithographs of M.C. Escher, or whatever Gardner's newly popularized subject *de mois* happened to be. In his 70-odd books, Gardner tackled topics from poetry to general relativity, from stage magic to epistemology, and his fans included notables like W.H. Auden, Vladimir Nabokov, Noam Chomsky, Stephen Jay Gould, and Carl Sagan.

My own geek-culture accomplishment—a lengthy streak on the hit quiz show *Jeopardy!* in 2004—doesn't quite put me in the Auden/Chomsky/Sagan bracket, but it was a direct result of my childhood love for Martin Gardner, and my fascination with his endless stream of paperbacks collecting favorite curiosities and brainteasers. When I was 8 or 9, I stumbled across his million-selling book *The Annotated Alice* in an elementary school library, and down the rabbit-hole I fell. This enthusiastic exegesis of the puns, puzzles, and paradoxes in Lewis Carroll's beloved books (including no less than 23 possible answers to the Mad Hatter's famously unanswered riddle "Why is a raven like a writing-desk?") was my gateway drug to the entire Gardner oeuvre, which seemed suffused not just with a joy of learning, but a joy of *algorithm*, of taking problems apart to figure out how they worked. His writing was a Charles Atlas course for an unenthusiastic math student like me—I went from being a mathematical 97-pound weakling to a professional computer programmer and, eventually, a game show celebri-nerd. I would often stand behind my *Jeopardy!* podium taking apart a tricky clue to figure out its inner logic, the way Martin Gardner had taught me to, waiting for that blissful moment of what he would call "aha! insight."

When I first brought *The Annotated Alice* home, I discovered there were already dozens of Gardner's pop-science books on our family room shelves; as it turned out, my father was a longtime Martin Gardner fanboy as well. In fact, I don't think he's ever been prouder than when I told him I'd been asked to edit a collection of Gardner's wordplay. Six months on *Jeopardy!*, delivering a top ten list for David Letterman, lecturing at the Smithsonian—these things were all well and good, but they paled next to the star power of Martin Gardner. "Will the book have Dr. Matrix in it?" he asked breathlessly, referring to Gardner's numerologically inclined alter ego from a series of *Scientific American* columns. (Dad will be relieved to hear that the good doctor makes a cameo appearance on page 52.)

My favorite Gardner columns were always those in which he dabbled in language and wordplay, so it was an unalloyed pleasure to revisit some of his favorites as we put together this book, which would sadly be his last. Martin Gardner died in Norman, Oklahoma, on May 22, 2010, while we were in the final stages of editing. He was 95 years old, and perhaps he was the last of the Renaissance men. His omnivorous and tirelessly curious intellect made him a throwback to a time when a gifted amateur (his spiritual ancestor Lewis Carroll, for example) could discover important insights in any field under the sun, or all fields all at once. It's possible that his passing marks the end of the era of the generalist, but I hope that's not so. Even in our time of credentialism and overspecialization, I hope there's still room for polymaths like Gardner. His ability to popularize math for the layperson was a result of the weeks of research he put into every column— "It took me so long to understand what I was writing about, that I knew how to write about it so most readers would understand it," he said—and he did it all without Google. Today's information technology could make us all into Martin Gardners, by making the sum of the world's knowledge more widely available than ever before. It's up to us to dive in—not just to study, but to *play*.

Remarkably, Martin Gardner never studied math formally after graduating from a Tulsa high school in 1932, but he did go on to earn a degree in philosophy at the University of Chicago and wrote extensively about philosophy and theology. He was famous as a skeptic of all kinds of superstition and pseudoscience, but, though his religious beliefs changed and developed over time, he always cheerfully believed ("by a hope and a leap of faith," he said) in an afterlife. So it's pleasant to imagine that he and Lewis Carroll might be good friends now, tackling some thorny chess problem or trading "word ladders" (see page 56) in the world to come.

Maybe he'll be able to find out, once and for all, why a raven is like a writing-desk.

—Ken Jennings
June 2010

INTRODUCTION

Most mathematicians and physicists that I have had the privilege of knowing have been fond of wordplay. It is not hard to understand why. After all, playing with words and letters is a whimsical form of combinatorics. Instead of numbers and other mathematical objects, the fundamental units of such play are letters. Palindromes, for example, can be based on both numbers and letters. Anagrams, pangrams, lipograms, and many other forms of wordplay have their analogs in number recreations.

Is finding a 10-digit prime that contains all 10 digits much different from constructing a sentence with just the 26 letters of the alphabet? Is an antimagic square (in which no two sums of the rows, columns, and diagonals are alike) much different from a word square in which every horizontal and vertical row is a different word?

Over the decades I have been almost as amazed by elegant wordplay as by elegant recreational mathematics. It occurred to me that it might be worthwhile to gather in a single volume some of my favorite word puzzles and other linguistic curiosities that have come my way. Many of these short items are original, but others are borrowed from here and there and are anonymous in origin. I have done my best to avoid much overlapping with other book collections of similar material.

I hope you enjoy playing along!

—Martin Gardner

1
Palindromes, Anagrams, and Other Whimsies

Palindromes

Palindromes are words, phrases, and sentences that read the same in both directions. It has been conjectured that when Adam and Eve first met, Adam said, "Madam, in Eden, I'm Adam." She replied, "Eve," while the serpent in the tree commented, "Tut, tut."

An owl in August, on the other hand, might complain, "Too hot to hoot!"

Thousands of clever palindromes have been constructed. Here are some of my favorites:

> Was it a car or a cat I saw?
> Draw pupil's lip upward.
> Ten animals I slam in a net.
> Poor Dan is in a droop.
> No, it is open on one position.
> Egad! A base tone denotes a bad age.
> Now, sir, a war is never even—sir, a war is won!

I once imagined a palindromic family consisting of *Pop* (*Blake DeKalb*), *Mom* (*Norah Sharon*), *Bob*, *Sis*, and *Tot*. They have a *pup* named *Otto* and drive *a Toyota*.

The word *radar* was coined as an acronym for *radio detection and ranging*, but choosing a palindrome for the new word was an elegant choice, since it suggests a signal going out and coming back.

Three of the finest palindromes in English were composed by Londoner Leigh Mercer:

> Straw? No, too stupid a fad. I put soot on warts.
> Sums are not set as a test on Erasmus.
> A man, a plan, a canal—Panama!

That third sentence, first published by Mercer in 1948, is perhaps the most famous and apt palindrome in history. Decades later, it spawned the following classic of "near miss" palindromes:

> A man, a plan, a canal—Suez!

Guy Jacobson expanded Mercer's famous palindrome as follows: "A man, a plan, a cat, a ham, a yak, a yam, a hat, a canal—Panama!" Peter Norvig, Google's Director of Research, has used a computer program to further expand the Panama palindrome into a 17,826-word version, which he calls the world's longest palindrome, though admittedly not a very interesting one.

Not every long palindrome results in incomprehensibility, however, as demonstrated by these gems:

> No, I save on final perusal—a sure plan, if no
>   evasion.
> Are we not drawn onward, we few, drawn
>   onward to new era?
> Doc, note. I dissent. A fast never prevents a
>   fatness. I diet on cod.

T. Eliot, top bard, notes putrid tang emanating,
    is sad. I'd assign it a name: gnat dirt upset on
    drab pot-toilet.

Palindromic novels have been published in both English and French, though such fiction makes very little sense. Laurence Levine's *Dr. Awkward & Olson in Oslo* and David Stephens's *Satire: Veritas* are among the classics of this dubious genre. Palindromic poetry has fared a little better, however. Poet Mike Maguire's memorable compositions include "Sun I Saw Was in Us," which begins:

> Sun in us.
> A sun I met.
> A faded light.

And ends:

> A path gilded.
> A fate minus
> A sun in us.

More of Maguire's palindrome verse is collected in his book *Drawn Inward and Other Poems*.

The hard-to-pronounce name *semordnilap* has been proposed for words that spell a *different* word when read backward. Some long examples in English include *lamina*, *reknits*, *stressed*, *samaroid*, and *rewarder*. Here's one interesting case: can you think of an English word that spells its own French translation (in plural form) when spelled backward?

A palindromic word square is one with words that are the reverse of their symmetrically opposite words. Such squares of side 4 are easy to construct. See the next page for an example:

```
R A T S
A B U T
T U B A
S T A R
```

Such squares have been made with sides 5 and 6, but none with common English words.

Other languages have palindromes of their own. In German: "Nie fragt sie: ist gefegt? Sie ist gar fein." (She never asks, "Is the sweeping done?" She is very refined.) Or in Spanish: "La ruta nos aportó otro paso natural." (The route provided us with another natural passage.) Speakers of Finnish can claim the longest single-word palindrome in common use, the nineteen-letter *saippuakivikauppias*, meaning "soap-stone vendor." By contrast, the longest palindromic words in English dictionaries tend to be iffy constructions of only eleven or twelve letters—*tattarrattat*, *kinnikinnik*, or *detartrated*, depending on the dictionary.

Some languages even *are* themselves palindromes. Thirty-six million people in southern India speak *Malayalam*, but only a few thousand residents of the impoverished Pacific island of Nauru still speak *Nauruan*.

Many long words are near-palindromes—that is, very close to reading the same forward and backward. The twelve-letter *sensuousness*, for example, would be a perfect palindrome if not for that pesky extra *s* at the end. *Footstool* would be a palindrome if only one letter were changed. Can you think of a vegetable whose nine-letter name is also one letter away from being a palindrome?

Words, rather than letters, can also be the units of a palindrome. Here are five clever examples, the first three by England's J.A. Lindon:

> You can cage a swallow, can't you, but you can't swallow a cage, can you?

King, are you sorry you are king?
Girl bathing on Bikini, eyeing boy, finds boy
eyeing bikini on bathing girl.
Fall leaves after leaves fall.
Blessed are they that believe that they are
blessed.

TRIPLE PUNS

A father of three sons bought a cattle ranch in Texas. When he asked his sons to suggest a name for the ranch they came up with *Focus*. Why? Because, they said, it's "where the sun's rays meet," a remarkable triple pun on "where the sons raise meat."

I found the following joke involving a triplet of great puns in a 2005 letter from Joe Miller in *Gilbert Magazine* (devoted to G.K. Chesterton). A thief stole several paintings from the Louvre. The Paris police captured him two blocks away when his van ran out of gas. Asked why, after such a well-planned crime, he stupidly failed to fill his car's gas tank, he replied "Monsieur, I had not the *Monet* to buy *Degas* to make my *van Gogh*."

And speaking of French art—did you hear about the celebrated painter who caused a terrible train wreck in Paris? It was Tooloose Latrack.

LIPOGRAMS

Word players have amused themselves by writing long paragraphs, poems, even novels, in which a certain letter is omitted. Such compositions are called lipograms, from the Greek for "missing symbol." The American novel *Gadsby*, written and self-published by Ernest Vincent Wright in 1939, does not contain a single *e*, the most frequently occuring letter in English. Wright produced his masterpiece by tying down the *E* key on his typewriter to ensure he didn't include one accidentally. The book is now a rarity.

In France, where *e* is also the most frequently occurring letter, Georges Perec, a famous novelist of the experimental "Oulipo" school, constructed a similar novel also avoiding the letter *e*. It is titled *La disparition* (*The Disappearance*). It is so skillfully composed that several French reviewers praised the novel without noticing the absence of *e*'s. *A Void*, Gilbert Adair's translation of the book into English, also has a vanishing *e*, and so does *Time* magazine's review of February 6, 1995. The translation of Perec's novel into Spanish, however, omits the letter *a*.

Perec later wrote another work, *Les Revenentes* (1972) in which *e* is the *only* vowel! The novella was later translated into English as *The Exeter Text: Jewels, Secrets, Sex*.

Stanford University's website reprints in full the previously unpublished text of "Autoportrait With Constraint," a 2008 address by author Douglas Hofstadter in which he tells his life story without using the letter *e*. It begins, "I was born in midtown Manhattan, right as World War Two was drawing to a, uhmm, to a *conclusion*." Hofstadter has written entire sections of his best-selling books in this *e*-less version of English he calls "Anglo-Saxon," since "English" itself starts with the forbidden letter.

Some lipograms are accidental. Take the sad case of Old Mother Hubbard:

> Old Mother Hubbard
> Went to the cupboard
> To get her poor dog a bone,
> But when she got there,
> The cupboard was bare,
> And so her poor dog had none.

Although I am sure it was unintended, this nursery rhyme does not contain the letter *i*. (Incidentally, speaking of letters, when Mother Hubbard opened the cupboard she might have exclaimed, "OICURMT!"—that is, "Oh, I see you are empty!")

Other lipogrammatic nursery rhymes are certainly intentional. My friend Ross Eckler rewrote "Mary Had a Little Lamb" in five different ways, each time omitting a different common letter. (The *a*-less version, for example, is called "Polly Owned One Little Sheep.") His *pièce de résistance* was a version of the rhyme that omits *thirteen* different letters—fully half the alphabet!

> Maria had a little sheep,
> As pale as rime its hair,
> And all the places Maria came
> The sheep did tail her there;
> In Maria's class it came at last,
> A sheep can't enter there;
> It made the children clap their hands—
> A sheep in class, that's rare.

An even more constrained collection of lipograms is *Eunoia*, the top-selling book of poetry in Britain for much of 2008. *Eunoia*, meaning "mental health," is the shortest word in the English language containing all five vowels, but the five sections in *Eunoia*, by Canadian poet Christian Bök, use only one vowel apiece. The fourth chapter, for example, is dedicated "To Yoko Ono," and begins, "Loops on bold fonts now form lots of words for books. Books form cocoons of comfort—tombs to hold bookworms."

Lipogram fans may also want to read James Thurber's children's classic *The Wonderful O*, a story about the banning of the letter *o* from the island of Ooroo. (The pirate captain responsible, whose ship is called the *Aieu*, has hated the letter *o* ever since he was a little boy and his mother was killed by an *o*-shaped porthole.)

DAFT DEFINITIONS

Avoidable: what a matador tries his best to do.

Hamnesiac: An actor who forgets his lines.

UCLA: What happens when the smog lifts in southern California.

Outwit: Dating. As in, "Who was that lady I seen you *outwit* last night?"

Hangover: The wrath of grapes.

Freudian slip: When you say one thing but mean your mother.

Décor: De part of de apple you don't eat.

Acupuncture: A jab well done.

Substitution clauses: Santa's elves.

Déjà moo: The feeling that you've heard this bull before.

PANGRAMS

The opposite of a lipogram is a pangram, a sentence that contains *all* letters of the alphabet. A perfect pangram would be a sensible English sentence that contained all 26 letters once exactly, but all the best attempts either rely on plentiful abbreviations:

TV quiz jock, Mr. Ph.D., bags few lynx.

or contain unfamiliar words and strained meanings:

Cwm fjord bank glyphs vex't quiz.

That last example is from Dmitri Borgmann, who notes that a *cwm* is a circular valley, *quiz* is an eighteenth-century term for an eccentric, and a *glyph* is a carved figure. So his pangram thus states than an eccentric person was annoyed by carved figures on the bank of a fjord in a circular valley. A near miss, at 27 letters, is this fairly straightforward sentence: "Big fjords vex quick waltz nymph."

Well-known imperfect pangrams include "A quick brown fox jumps over the lazy dog" (33 letters), "Pack my box with five dozen liquor jugs" (32 letters), and "Jackdaws love my big sphinx of quartz" (31 letters). Sentences like these are often used as sample text in word processors and other computer programs, since they can smoothly display all the letters in a particular typeface.

If a perfect pangram were expressed as a cryptogram, without spacing between words, it might look like this:

ABCDEFGH IJKLMNOPQRSTUVWXYZ

The artificial nature of pangrams has led some devotees to search for "pangrammatic windows"—short stretches of naturally occurring text that contain all 26 letters of the alphabet, not by design but by coincidence. An *Entertainment Weekly* news item in February 2006 contained a remarkably short 61-character window, indicated here in bold italics:

> On Jan. 26, [director Werner] Her**zog happened by Joaquin Phoenix's car wreck and pulled the actor from the** vehicle. Somebody hook this brother up with Jerry Bruckheimer.

The shortest known window in published English text is a run of 56 letters from *In the Courts of Memory*, a 1912 memoir by the diplomat's wife Lillie de Hegermann-Lindencrone. ("... I thought I san**g very well but he just looked up into my face with a very quizzical ex**pression ...") Note that even the shortest pangrammatic window is over twice the length of the shortest artificial pangram.

WORD SUPERLATIVES

What's the longest word in the dictionary? There are many joke answers. *Rubber,* because you can stretch it. *Smiles,* because there's a mile between the first and last letters. *Beleaguered,* because there's a league between the beginning and the end. *Endless,* because there's no end to it.

If hyphenated words are allowed, there *is* no longest word, since one can speak of a great-great-great-...grandmother. But if we ignore hyphenated words, as well as made-up words, chemical compound names, medical terms, surnames, place names, and other artificially constructed words, the longest common words in English are such 22-letter words as *deinstitutionalization* and *counterrevolutionaries.* The longest dictionary word (aside from the exceptions noted above) is still the 28-letter *antidisestablishmentarianism,* which can be lengthened by replacing *-ism* with *-istically.*

The two best-known long invented words are Shakespeare's *honorificabilitudinitatibus* (spoken by Costard the clown in *Love's Labour's Lost*) and Mary Poppins's *supercalifragilisticexpialidocious.* Do you notice something else unusual about the construction of Shakespeare's word?

The longest word ever to appear in an English dictionary is *pneumonoultramicroscopicsilicovolcanoconiosis,* which refers to a lung disease caused by breathing in volcanic dust. The word can be found in the Third Edition of Merriam-Webster's Unabridged as well as the Oxford English Dictionary, though it's never actually been used by the medical field. It was invented as a stunt by Everett M. Smith at a 1935 meeting of the National Puzzlers' League, of which Smith was then president. Members of the League subsequently mounted a successful campaign to get the word recognized by dictionaries.

It is often asserted that the longest word in the Oxford English Dictionary is *floccinaucinihilipilification*, meaning the action of estimating as worthless, but the word is usually spelled with four hyphens.

Though it's also sometimes hyphenated, the longest word in the Bible is the 18-letter *Mahershalalhashbaz*, meaning "quick to the plunder, swift to the spoil." The prophet Isaiah is instructed to give his son this jawbreaking name in Isaiah 8:3.

Strengths is a one-syllable word with nine letters. *Scraunched, squirreled,* and *strengthed* each have 10 letters. *Broughammed* (discovered by William Harmon) is longer still with 11 letters!

Aegilops, a genus of plants called goatgrasses, is the longest word with all its letters in alphabetical order. More common alphabetical words are six-letter entries like *almost, biopsy,* and *chintz,* but if repeat letters are allowed, *billowy* has the same property. *Wronged* and *sponged* both have their letters in reverse alphabetical order. *Spoon-feed* and *trollied* are longer, but use repeated letters. *Decontextualizations* alternates steps forward in the alphabet with steps backward.

Uncopyrightables has 16 letters with no letter appearing twice.

Defenselessnesses and *strengthlessnesses* are the longest words in which only a single vowel is used. The longest words to use only a single vowel that's not the letter *e* are 13-letter words with all *i*'s, like *primitivistic* or *philistinisms.*

Typewriter, by a surprising coincidence, can be typed with only the top row of letters on a typewriter. Other long words with the same property are *perpetuity* and *repertoire. Alfalfa* uses only the typewriter's middle row of letters. No word can be typed with the bottom row because it contains no vowel or *y.*

The letters in the word *stewardesses* can be typed with only the left hand. Even longer possibilities include *sweaterdresses, street addresses*, and *Babette's Feast* (well, except for the apostrophe). *Honolulu*, *polyphony*, and *opinion poll* are possible using only the right hand. Among U.S. presidents, *Polk* is the only one whose name can be typed with the right hand. Only *Taft* and *Carter* are typed with the left hand. Among states, only *Ohio* can be typed with the right hand, only *Texas* with the left.

The only one-syllable state, incidentally, is *Maine*.

Blackballed and *blackmailed* are the longest common words that use only letters in the alphabet's first half. Other interesting phrases that qualify are *Cecil B. DeMille*, *Black Like Me*, *baggage claim*, and *fiddle-de-dee*. *Soupspoons* is the longest common word from the second half of the alphabet; you can also make *topsy-turvy* and *not to worry*.

Ambidextrous is a most remarkable word. Its first six letters are all in the first half of the alphabet, and its last six letters are all in the second half. It is the longest known example of what the November 2001 issue of *Word Ways* calls an ambidextrous word—one with its first half in the alphabet's first half, and its second half in the alphabet's second half. Such words must have an even number of letters to avoid the problem of a middle letter, and the word should contain no duplicate letters.

No ambidextrous common word of 14 letters has been found, and the same goes for common antidextrous words (in which the first half of the word contains letters from the last half of the alphabet, and vice versa). *Unprovidable* is an example of a 12-letter antidextrous word.

Angsts is the shortest word with five consecutive constants.

Facetiously and *abstemiously* have all six vowels, including *y*, in alphabetical order.

Taking *y* as a vowel, *aye* has no consonant. An *aye-aye* is a nocturnal lemur. The vowel-less *euouae* is a sequence of tones in medieval chants.

Pizzazz has four *z*'s. *Knickknack* has four *k*'s.

Banana, cocoon, pepper, googol, horror, and *mammal* are among the many words with the property that one letter is used once, another letter twice, and a third letter three times. Among words that extend the property to a fourth letter that appears four times are *rememberer* and *sleeveless.*

In a common cipher, letters are represented by their mirror-image counterparts from the other half of the alphabet. Thus *A* is coded by *Z*, *B* by *Y*, *C* by *X*, and so on. A pleasant pastime is to search for words that, when coded, become other words, like *girl* to *trio*, or *grogs* to *tilth*. Can you find a common four-letter word that codes to a number in this cipher? Careful now.

A symmetric word is one that would be represented, in this cipher, by its own reversal. *By* would be *yb*, for example, and *grit* would be *tirg*. The longest words with this property are six-letter words: *wizard* and *hovels.*

The two-syllable *rugged* shortens to one syllable when *sh* is added to make *shrugged*. The word *pat*, when spelled in reverse, has the same meaning. I found both these oddities in Owen O'Shea's delightful book of coincidences *The Magic Numbers of the Professor* (2006).

Dreamt is the only common English word ending in *mt*. *Asthma* is the longest common English word that has a vowel at the beginning and end, and only consonants in between. Less common examples include *isthmi* (more than one isthmus) and *archspy*. And *rhythms, nymphly,* and *tsktsks* are the longest words that contains none of the five vowels.

Many common English words have a double *i* in them: *radii, shiitake, skiing, taxiing*. But there was no triple-*i* word until 2007, when *Wiiitis* was coined. A letter in the *New England Journal of Medicine* gave this name to a medical condition similar to tennis elbow, but caused by playing tennis on a Nintendo Wii.

Usher is unusual in containing, in consecutive letters, four pronoun words: *he, she, her,* and *us*. *Therein* may be the shortest word that contains within it, in the same fashion, 11 other common words. (Disregard single-letter words.) Can you find them all?

In the April 22, 1979, *Philadelphia Inquirer*, Theodore Bernstein quoted Leo G. Staley of Columbus, Ohio, as finding 10 common words in *scapegoat*, and Ralph Beaman subsequently discovered that *firestone* was an even better nine-letter choice, since it contains 13 common words. In a similar vein, Boris Randolph of Los Angeles discovered 19 words in *misinformation*. How many words can you find in *scapegoat*, *firestone*, and *misinformation*?

LINGUISTIC CATCHES

Ask someone how to pronounce the three words *to*, *two*, and *too*. Then ask, "How do you pronounce the second day of the week?" If your victim says "Tuesday," you say, "That's strange. I always thought it was Monday."

"Would you like to have your palm read?" "Sure." Produce a stick of red lipstick and smear some on your victim's palm.

Say to someone, "You are driving a bus. It contains seven people: four men and three women. At the next stop two men get off, and one man gets on. At the next stop, five women get on." Continue, giving details about who gets on and who gets off. Finally say, "Now tell me the name of the bus driver." The victim is almost sure to forget that he or she is the driver!

What book has its preface in the middle, its appendix at the front, and reaches a climax before you ever come to the plot? The dictionary, of course.

If you check the second edition of *Webster's New International Dictionary* (1934, page 771), you'll find listed the curious word *dord*, which is defined as "density." This nonexistent word crept into the dictionary by mistake when someone misinterpreted a memo that listed "D or d" as possible abbreviations for "density." When the error was finally caught, five years later, the plate was quickly changed, and a nearby entry for "Dore furnace" was expanded to fill in the extra space. Editor-in-chief Philip Babcock Gove wrote that the loss was "probably too bad, for why shouldn't *dord* mean density?"

The *New Oxford American Dictionary* also contains a nonexistent word, but this one was included on purpose. The editors invented the word *esquivalience*, claiming that it means "the willful avoidance of one's official responsibilities." Why the false word? To trap any other dictionary that might try to steal entries from the *New Oxford*. Editor-in-chief Erin McKean says this practice is "not uncommon" among lexicographers, though no other intentionally false dictionary words have yet been discovered.

The dictionary staff at Merriam-Webster says that it still receives about four letters a year demanding an answer to the notorious -*gry* puzzle. This is a riddle that first made the rounds in early 1975: "There are three common English words that end in the letters -*gry*. Two are *angry* and *hungry*. What is the third?"

Though this is probably the best-known word puzzle of all time, it's a complete hoax: there *is* no other common word ending in *gry*. Possible archaic answers found in various dictionaries include *aggry*, a kind of African glass bead, or *puggry*, a light Indian turban. Here's a similar riddle that *does* have an answer: can you think of two English words that end with *-shion*? I assure you that this isn't a trick question!

Finally, is there a synonym for *synonym*? Yes, the *Oxford English Dictionary* lists it as *polyonym*. *Polyonymous*, according to *Webster's College Dictionary*, means "having or known by various names."

INVERSIONS

Students of curious wordplay have long recognized that short words can be formed to display various types of geometrical symmetry. Upside down *W* turns into *M*, so COW inverted is almost *MOO*. Is *MOM* beautiful? Upside down, WOW! I have been in tall buildings in which each men's room was indicated by a large brass M on the door, and each ladies' room by the same fixture installed upside down. Has anyone ever used a similar inversion device next to the buttons of an elevator: "up" and "dn"?

Other words printed in block letters, such as NOON, are the same upside down. A remarkable example of a sentence, said to be on a sign near a swimming pool that was newly closed on Mondays, is unchanged when rotated 180 degrees:

NOW NO SWIMS ON MON

When uncapitalized, *chump* is a rare word that has the same property in script:

chump

In Paris a clothing shop called "New Man" has a sign on which its logo is lettered "**NEW MƎN**," with the *e* and the *a* identical except for their orientation. As a result the entire sign has upside-down symmetry. The logos of *ZOONOOZ* (the magazine of the San Diego zoo), Nissin (a Japanese maker of ramen noodles), and Xpedx (a paper distributor) are all similarly designed to feature 180-degree rotational symmetry,  as is the logo of Edge Composites, a bicycle parts brand; that logo is readable no matter which part of the bicycle's tire is facing up. The United Nations Association's magazine, *Vista*, also once featured a rotational logo.

Other logos feature different sorts of ambigrammatic symmetry. Sun Microsystems' logo reads "SUN" in four different directions. The logo of the DeLorean Motor Company has left-right symmetry, mirrored across its center axis.

One day in a supermarket my sister was puzzled by the name on a box of crackers, "spep oop," until she realized that a box of "doo dads" was on the shelf upside down. Wallace Lee, a magician in North Carolina, liked to amuse friends by asking if they had ever eaten any "ittaybeds," a word he printed on a piece of paper like this:

Ittaybeds

After everyone said no, he would add: "Of course, they taste much better upside down."

Other words have mirror symmetry about a vertical axis, such as "bid" (and "pig" if the *g* is drawn as a mirror image of the *p*), and some have rotational symmetry. At the top of the next page you'll see an amusing way to write "minimum" so that it is the same when it is rotated 180 degrees:

It is the gifted author and puzzle creator Scott Kim who has carried this curious art of symmetrical calligraphy to heights not previously known to be possible. By ingeniously distorting letters, yet never so violently that one cannot recognize a word or phrase, Kim has produced incredibly fantastic patterns. His 1981 book *Inversions* is a collection of such wonders, interspersed with provocative observations on the nature of symmetry, its philosophical aspects, and its embodiment in art and music as well as in wordplay. Consider Kim's lettering of my name below. Turn it upside down and presto! It remains exactly the same.

For several years Kim's talent for lettering words to give them unexpected symmetries was confined to amusing friends and designing family Christmas cards. He would meet a stranger at a party, learn his or her name, then vanish for a little while and return with the name neatly drawn so that it would be the same upside down. His 1977 Christmas card featured upside-down symmetry, with the four recipients of the card also appearing as individual ambigrams. (Lester and Pearl are his father and mother; Grant and Gail are his brother and sister.)

The next year he found a way to make "Merry Christmas" mirror-symmetrical about a horizontal axis, and in 1979 he made the mirror axis vertical. Those designs appear on the next page.

For a wedding anniversary of his parents Kim designed a cake with chocolate and vanilla frosting in the pattern shown at right. ("Lester" is in black, "Pearl" is upside down in white.) This is Kim's "figure and ground" technique. You will find another example of it in *Gödel, Escher, Bach: An Eternal Golden Braid*, the Pulitzer-winning book by Kim's good friend Douglas R. Hofstadter.

Here Kim has lettered the entire alphabet in such a way that the total pattern has left-right symmetry.

All the patterns in Kim's book are his own. A small selection of a few more is given below to convey some notion of the amazing variety of visual tricks Kim has up his sleeve.

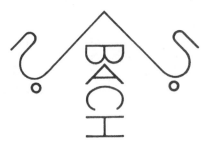

MEDICAL MIRTH

Party guest: May I ask what you do for a living?
Stranger: I'm a naval surgeon.
Party guest: Goodness me, how you doctors specialize!

Patient (after a physical examination): Well, doctor, how do I stand?
Doctor: That's what puzzles me.

Psychiatrist: Do you have trouble making up your mind?
Patient: Well, yes and no.

Secretary: There's an invisible man in the waiting room.
Doctor: Tell him I can't see him.

WORD SQUARES

Word squares are a popular form of wordplay that goes back to ancient times. The idea is to fill the cells of an n-by-n matrix with letters that spell a different word in each horizontal row, and with the same words in the columns. The more familiar the words, the more elegant the square. Such squares obviously become more difficult to construct as n increases. There is a vast literature on this pastime in all major languages. Many amazing word squares have been created for $n = 9$, but so far none of order 10 has been found that is not marred by one or more strange words.

In 1989, Eric Albert used his home computer to find the first nine-square using only words from a single dictionary. His remarkable square, still one of the best of its kind, is presented on the next page.

N E C E S S I S M
E X I S T E N C E
C I R C U M F E R
E S C A R P I N G
S T U R N I D A E
S E M P I T E R N
I N F I D E L I C
S C E N A R I Z E
M E R G E N C E S

See how long it takes you to square the circle and the cube by solving the two samples that follow.

¹C	²I	³R	⁴C	⁵L	⁶E
²					
³					
⁴					
⁵					
⁶					

2. Accustoms
3. Most insulting
4. Fold
5. Smaller
6. Aromatic compounds

¹C	²U	³B	⁴E
²			
³			
⁴			

2. Not pretty
3. Feeling sad
4. Ogles

No good solutions are known for squaring *square* or *triangle* using traditional word square rules (with no phrases and no proper names). Bending the rules to allow more possible entries makes squaring the square, at least, a simple matter:

¹S	²Q	³U	⁴A	⁵R	⁶E
2					
3					
4					
5					
6					

2. Australian airline
3. Turmoil
4. "Relax, troops!"
5. Fight, country-style
6. Think highly of

For cubing *cube* and similar tasks see page 397 of my *Colossal Book of Short Puzzles and Problems* (Norton, 2006).

It is believed to be impossible to place nine letters, no two alike, inside a 3×3 matrix so that a familiar word is spelled in eight different ways: rows, columns, and diagonals. (Ross Eckler has managed several, but all rely on using the French word *eau*.) There are thousands of near misses in which one letter is duplicated. Here is a typical example with one duplicate letter. A clue for the top row is "Uncle _____," and one for the middle row is "Cherry _____." This should be enough information to enable you to construct the square.

Mathematician Tom M. Apostol's secretary Maria Suzuki challenged him to find a 3×4 word rectangle with a common word on each row, a common word on each column, with no letters duplicated, and that included all five vowels plus *y*. Apostol found the amazing solution shown below. Perhaps you can find others?

S A I D
U G L Y
P O K E

RIDICULOUS RIDDLES

1. What does a duck do when it flies upside down?
2. If everyone in America drove a pink Cadillac, what would the country be?
3. What did George Washington say to his men before they got in their boats to cross the Delaware?
4. What do you get when you cross a hippo with a jar of peanut butter?
5. What's the smartest state in the U.S.A.?
6. Why do baby ducks walk softly?
7. What do you get when you drop a bomb on a French kitchen?
8. What is green and flies through the air?
9. Who was the tallest president of the United States?
10. What has a hump, is brown, and lives at the North Pole?

CHOP WORDS AND SNOWBALL SENTENCES

A chop word is a word with the following property: By removing some letter from within the word, you can always leave another word until you are down to a single letter such as *a* or *I*. *Restarted* is an example of a common English chop word of nine letters. Successive chopping leaves *restated*, *restate*, *estate*, *state*, *stat*, *sat*, *at*, and *a*.

There are many other nine letter chop words, such as *startling*, *cleansers*, *sparkling*, and *stampeded*, but is there a longer common word with this property? Proper names are not allowed. If you can find such a word with ten or more letters, send a note in care of this book's publisher.

A snowball sentence, on the other hand, is one that gets *longer* as it goes, like a snowball rolling down a mountain—in fact, each of its words must be one letter longer than its predecessor. Harry Mathews, the only American member of the French "Oulipo" school, gave a French example of 22 letters, beginning with *O le bon sens* and ending with *pseudotransfigurations*. A good English specimen, from Dmitri Borgmann's classic *Language on Vacation*, is the following 20-word snowball:

> I do not know where family doctors acquired illegibly perplexing handwriting; nevertheless, extraordinary pharmaceutical intellectuality, counterbalancing indecipherability, transcendentalizes intercommunications' incomprehensibleness.

DECAPITATIONS AND CURTAILMENTS

Do you notice anything remarkable about the following passage? "Show this bold Prussian that praises slaughter, slaughter brings rout. Teach this slaughter lover his fall nears." If each word is beheaded—that is, has its first letter removed—two entirely new sentences result.

George Canning, an early 19th-century British statesman, wrote the following verse about a word that is subject to "curtailment," that is, a word that becomes a different word when its last letter is removed. Can you identify the word?

> A word there is of plural number,
> Foe to ease and tranquil slumber;
> Any other word you take
> And add an "s" will plural make.
> But if you add an "s" to this,
> So strange the metamorphosis,
> Plural is plural now no more,
> And sweet what bitter was before.

Both decapitation and curtailment are involved in the following old riddle:

> From a number that's odd,
> Cut off the head,
> It then will even be;
> Its tail I pray now take away,
> Your mother then you'll see.

SHIFTWORDS

L. Frank Baum set his classic series of children's books in a magical land called Oz. Shift each letter of the word *OZ* back one step in the alphabet and you get *NY*, which was Baum's home state. Shift each letter of *OZ* forward one step (*Z* wrapping around to *A*) and you get *PA*, the home state of Ruth Plumly Thompson, the author who continued the Oz series after Baum's death.

Shiftwords are word pairs like *NY* and *OZ* that can be translated into one another via uniform movement through the alphabet. Take the word *adds*, for example. Move each letter in *adds* once step forward in the alphabet, and the result is *beet*.

Not all shiftword pairs involve a single step in the alphabet. The letters in *cold*, for example, must each be translated *three* steps forward to produce *frog*. *Pecan* is four steps away from *tiger*. *Abjurer* is thirteen steps away—as far as you can get—from *nowhere*. (As with *OZ* and *PA* above, shifts past the end of the alphabet start over again with *A*.)

Abjurer is a somewhat uncommon word, as is *primero*, which is three steps back from *sulphur*. There doesn't seem to be a seven-letter shiftword pair in English that uses common words. Good six-letter examples include *steeds-tuffet* and *fusion-layout*.

The most memorable shiftword pairs are those whose meanings seem to be related in some way. *Irk* and *vex*, though a diametrically opposed 13 steps apart, are exact synonyms. *Jolly* and *cheer* are very closely related, as are *green* and *terra*. And bilingual word enthusiasts will be *enchanté* to learn that *yes* can be shifted back 10 steps to make *oui*.

WIT FROM THE HEADLINES

After being sworn in as vice president, Gerald Ford displayed a knack for self-deprecating automotive humor. "I'm a Ford," he told reporters, "not a Lincoln."

Bill Clinton, while stumping for his wife during her presidential campaign, said he looked forward to becoming the country's "first laddie."

In his eulogy at Ronald Reagan's funeral, George H.W. Bush recalled the former president being asked how a meeting had gone with Archbishop Tutu. Reagan couldn't resist replying, "So-so."

In a *New York Times* column, William Safire once called Hillary Clinton a "congenital liar." Many years later, on a talk show, he calmed the waters by saying he had meant to call her a "congenial lawyer."

ANAGRAMS

An anagram is the rearrangement of the letters in one word or phrase to form another, a pastime that dates back at least to the ancient Greeks. *Enumerations* is an anagram of *mountaineers*. *Pictures* is an anagram of *piecrust*. Though there is no one-word anagram for the word *anagrams*, many anagrammatists refer to their pursuit as "Ars Magna," Latin for "Great Art."

Famous anagram fans from history include Plato, Galileo, and even King Louis XIII of France, who paid Thomas Billon, a Provençal subject, 1,200 livres a year to serve as his Royal Anagrammatist. A number of authors have used anagrams of their names to settle on noms de plume: François Rabelais published his first book, *Pantagruel*, under the anagrammatic pen name "Alcofribas Nasier." Vladimir Nabokov has used the name "Vivian Darkbloom," and Edward Gorey published books as Ogdred Weary, Regera Dowdy, and many, many others. Honoré de Balzac signed his first three novels using the surname "R'Hoone." Most scholars believed that François-Marie Arouet took his famous pseudonym, Voltaire, by anagramming the phrase "Arouet, l.j." (*le jeune*, or "the younger") and replacing *u* with *v* and *i* with *j*, as was common at the time.

In the early 20th century, a series of articles on fish and reptiles written by one "H.A. Largelamb" appeared in *National Geographic* magazine. In fact, these articles were written by a great American scientist and engineer, who used a pen name because he suspected that magazines were accepting his work only because of his fame. Can you deduce H.A. Largelamb's true identity?

John Donne even wrote an elegy called "The Anagram," which contains the following couplet:

> Though all her parts be not in th' usual place,
> She hath yet an anagram of a good face.

Norman Mailer, in his 1973 book about Marilyn Monroe, went out of his way to point out that if the *A* in *Marilyn* is used twice, the *O* in *Monroe* used just once, and the *Y* omitted, the remaining letters can be arranged to spell *Norman Mailer*. No one can deny that Norman and Marilyn were very close.

The Wall Street Journal reported on September 6, 1978, that a Florida-based company called Xonex, which makes motor oil, was being sued by Exxon for using the letters of its name in anagram form. Patrik J. McEnary, president of Xonex, said it never occurred to him that the two names were anagrams until he received a letter from an Exxon attorney; in fact, he thought the letter was a hoax.

Lewis Carroll was fond of composing anagrams on the names of famous people. He anagrammed *William Ewart Gladstone* several ways, perhaps the best being "Wild agitator! Means well." Gladstone's political rival, *Disraeli* was anagrammed by Carroll as "I lead, sir!" He also composed "Flit on, cheering angel!" for *Florence Nightingale*.

Here are some surprising anagrams. Some, like the first entry, date back to the 18th century. Others are of more recent vintage.

astronomers	moon starers
the eyes	they see
a decimal point	I'm a dot in place
the Mona Lisa	no hat, a smile
desperation	a rope ends it
earthquakes	that queer shake
election results	lies, let's recount
George Bush	he bugs Gore
a shoplifter	has to pilfer
William Shakespeare	we all make his praise
halitosis	Lois has it
slot machine	cash lost in 'em
dormitory	dirty room

One of the greatest feats in anagram history is David Shulman's 1936 sonnet "Washington Crossing the Delaware," a poetic retelling of the title event in which every line of the poem is an anagram of the title.

Washington Crossing the Delaware

A hard, howling, tossing water scene.
Strong tide was washing hero clean.
"How cold!" Weather stings as in anger.
O silent night shows war ace danger!

The cold waters swashing on in rage.
Redcoats warn slow his hint engage.
When star general's action wish'd "Go!"
He saw his ragged continentals row.

Ah, he stands—sailor crew went going.
And so this general watches rowing.
He hastens—winter again grows cold.
A wet crew gain Hessian stronghold.

George can't lose war with's hands in;
He's astern—so go alight, crew, and win!

Shulman wrote to *The New York Times* in 1996, "Just as a magician does not like to disclose his modus operandi, so I am loath to disclose mine. . . . After waiting 60 years, I find that nobody so far has equaled or surpassed it. I even tried to, but I failed."

Dmitri Borgmann has proposed the name "antigrams" for words that can be anagrammed to the *opposites* of their plain meanings. Some examples:

Infection—fine tonic
Misfortune—it's more fun
Militarism—I limit arms
Evangelists—evil's agents

In an even more rarefied class of anagrams, no rescrambling of letters is necessary at all: the parallelism is achieved through changing the spacing of the word or phrase. An *island* "is land." Someone who is a *daredevil* might have "dared evil." And calling someone *gentlemanly* means that they're both "gentle" and "manly." Lewis Carroll might have enjoyed this kind of "anagram," since he once wrote a short story in which a man misreads a shop sign for "Roman Cement" and mistakenly thinks he can find *romancement* within.

"BETCHAS"

Bet someone you can ask three questions and he will answer "No" to at least one of them. Question one: Do horses have five legs? Of course he will answer yes. Question two: Do dogs meow? Again he will say yes. Now you say: "Wait, maybe we should stop. Have you heard this one before?" Caught off guard, he may reply, "No I haven't." If so, you win the bet!

Bet you can stay under water for five minutes. Hold a glass of water on top of your head.

Bet someone she can't write a lower case i and put a dot over it. She'll probably write "i" instead of this:

SPOONERISMS

The Reverend William Archibald Spooner was a beloved Oxford don for over sixty years, from 1867 to 1930. Today, he's mostly remembered for lending his name to "spoonerisms," his trademark gaffes in which nearby consonants, vowels, or entire syllables are accidentally switched, to humorous effect. No doubt many of these stories are apocryphal, but his students loved to recount catching Spooner in errors like calling Queen Victoria "our queer old dean" or referring to the Lord as "a shoving leopard." "Someone is occupewing my pie," he is said to have once told a church usher. "Please sew me to another sheet."

Twentieth-century spoonerist Bennett Cerf reminded us of the man who poured pickle juice down a hill to see if dill waters run steep, the woman who selected a paperback from the trite side of the racks, and the lucky baker who found a four-loaf cleaver. Clifton Fadiman pointed out that combined charity drives put all their begs in one ask-it. "Will you please hush my brat?" said the visitor to the butler. "It's roaring with pain outside." Is it in one of George Kaufman's plays that a coed is said to have put her heart before the course? And have you read *Lady Loverley's Chatter?*

The story may be apocryphal, but Adlai Stevenson is supposed to have been asked about his religious influences during the 1960 presidential campaign, after clergyman Norman Vincent Peale had made some unfortunate remarks about John F. Kennedy's Catholic beliefs. Stevenson told the press that he found St. Paul appealing and St. Peale appalling, surely one of the finest of all topical spoonerisms.

The great spoonerisms are more memorable than the clichés they garble. "I am a conscientious man—when I throw rocks at sea birds, I leave no tern unstoned," wrote Ogden Nash. And it was radio comedian Jane Ace who first observed that "time wounds all heels."

ANSWERS TO CHAPTER 1

PALINDROMES

The English word that spells its own French plural form backward is *state* (*etats*).

Red pepper is one letter (or one 180-degree turn of a letter) away from being a palindrome.

WORD SUPERLATIVES

Shakespeare's invented word, *honorificabilitudinitatibus*, alternates consonants with vowels throughout.

CORE, using a symmetric cipher, codes as XLIV, 44 in Roman numerals.

Therein contains *the, there, he, her, here, herein, er, ere, re, rein,* and *in. Scapegoat* contains *scape, cap, cape, ape, peg, ego, go, goat, oat,* and *at. Firestone* contains *fir, fire, fires, ire, ires, re, rest, stone, to, ton, tone, on,* and *one. Misinformation* contains *mi, mis, misinform, is, sin, in, inform, information, for, form, format, formation, or, ma, mat, at, ti, ion,* and *on.* Randolph puckishly noted that "The letter *M* is in *formation.*"

DICTIONARY DIVERSIONS

The only two common words that end with -*shion* are *cushion* and *fashion* (and derivatives thereof).

Word Squares

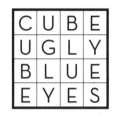

C	I	R	C	L	E
I	N	U	R	E	S
R	U	D	E	S	T
C	R	E	A	S	E
L	E	S	S	E	R
E	S	T	E	R	S

C	U	B	E
U	G	L	Y
B	L	U	E
E	Y	E	S

S	Q	U	A	R	E
Q	A	N	T	A	S
U	N	R	E	S	T
A	T	E	A	S	E
R	A	S	S	L	E
E	S	T	E	E	M

S	A	M
P	I	E
A	R	T

Ridiculous Riddles

1. Quack up
2. A pink car-nation
3. "Men, get in the boats!"
4. A 5,000-pound sandwich that sticks to the roof of your mouth
5. Alabama—it got four A's and one B
6. Because they can't walk at all hardly
7. Linoleum Blownapart
8. Super Pickle!
9. Dwight D. Eiffeltower
10. A lost camel

Decapitations and Curtailments

The answer to the first riddle is *cares*, which becomes *caress*. The answer to the second is *seven* (*even* and *Eve*).

Anagrams

"H.A. Largelamb" was an anagram for "A. Graham Bell"—telephone inventor Alexander Graham Bell.

2
Perplexing Puzzles and Curious Conundrums

1. Crazy Crossword

ACROSS
1. Pester
5. Exploit
6. Hive dwellers
7. Relaxation

DOWN
1. Television set
2. Toothpaste holder
3. Duct
4. London Underground

2. A Monster Word

I spotted this monstrous 21-letter word in a *New York Times* review (June 27, 2005) of a chamber music concert. I've filled in the blanks below with the vowels *E*, *I*, and *O*. Can you supply the missing consonants?

I _ _ O _ _ _ E _ E _ _ I _ I _ I _ I E _

3. Blind Bus Driver

Your friend tells you, "They've just hired a blind man to drive the school bus." How can this be true? If you interpret the sentence properly, it will make good sense.

4. Missing Words

In the blank squares shown above, see if you can fill in letters so that each column is a three-letter word, and the middle row spells the name of a shelled sea animal. There are two different solutions!

5. What Month?

What simple procedure will change these strange symbols to the name of a month?

6. Stop and Snap

"Now that we've done the dishes," said Mrs. Rendrag to her son Nitram, "Let's do the *stop* and *snap*."

Can you manipulate the two italicized words to find out what Mrs. Rendrag really suggested? If you see the trick, you'll have no difficulty learning the real name of the mother and her son.

7. Anna's List

Anna Graham is mailing holiday cards to eight friends, but she can't remember which U.S. state each lives in. By studying their names, can you deduce each person's home state?

Roy Kewn
Nora I. Charlton
Colin A. Fair
Dora K. Hatton
Earl Wade
Hilda D. Rosen
A.K. Barnes
J.R. Sweeney

8. In and Out

On the front of a building there is the word BANK in large letters above two doors. One door has on it the letters ENTER. The other door says EXIT.

Add a single letter to each of the three capitalized words in the previous paragraph to make a new word. The new words fit the blank spaces of the following three sentences:

1. Do unicorns _____?
2. He had a _____ expression on his face.
3. H is the _____ of nothing.

9. Space Oddity

In the Arthur C. Clarke novel and subsequent film *2001* there is a supercomputer named HAL. There is a curious relationship between the acronyms for HAL and the computer company IBM. Can you discover it?

10. Flying Saucers

A woman was asked if she had ever seen a UFO, or unidentified flying object. Remember the HAL-IBM relationship in the previous puzzle? *UFO* relates to a second three-letter word in much the same way. With that relationship in mind, how many UFOs did the woman say she had seen?

11. Two Little Words

We _____ on the _____way,
     a   b

And we _____ on the _____way.
       b   a

Put the same word in both the spaces marked *a* and put a different word in both the spaces marked *b* to make sense of the sentence.

12. Prepositional Logic

Students are sometimes told never to end sentences with a preposition, though in reality that's not a strict grammatical rule. (When this convention was pointed out to him, Winston Churchill is reported to have said, "This is the sort of English up with which I will not put!")

Humorous sentences have been composed to illustrate serious infractions of this "rule." One of the most famous ends with the five prepositions "to out of up for," in that order. Can you reconstruct that sentence?

13. A Peculiar Word

Can you think of a five-letter word with the following unusual property? If you remove four of its letters, what remains is pronounced exactly the same way as the original word.

CLASSIC CLASSROOM CATCHES

Ask a victim to recite several times the following mysterious words: *owa tagu siam*. It may take a while for him to realize he is saying "Oh what a goose I am!"

"If ice in water makes iced water, what does ice in ink make?" If a friend replies, "Iced ink," you respond, "In that case, you should bathe more often."

Ask a person to touch her head and say the abbreviation for *mountain*. If she touches her head and says "M-T," reply, "You said it!"

Touch a person's elbow or knee and say, "I'll bet you didn't know that an Eskimo doesn't have this particular bone." If he expresses doubt, say, "Of course, he has a bone just like it."

14. Odd vs. Even

All letters in odd positions on the pattern above are on gray cells. It is a remarkable coincidence that all six vowels, including *Y*, are on gray cells. But despite the lack of vowels, there is a three-letter word, often used by mathematicians, that can be spelled using three letters on the white cells. Can you find it?

15. WILLIAM'S STRANGE PREFERENCES

William likes apples better than oranges, and vanilla ice cream better than chocolate. He would rather watch a baseball or football game on TV than hockey, enjoys summer and fall more than spring and winter, considers *Newsweek* to be a better magazine than *Time*, and is convinced that *High Noon* is a greater Western movie than *Destry Rides Again*.

Can you explain why William has these preferences?

16. PERVERBS

In his book *Selected Declarations of Dependence* (Eternal Network, Toronto, 1976), author Harry Mathews explores hundreds of what he calls "perverbs," sentences formed by joining the first half of one proverb to the second half of another. Can you recite the opposite "perverb" of each of these? For example, the opposite of "People who live in glass houses are soon parted" would be "A fool and his money shouldn't throw stones."

1. "A stitch in time gets the worm."
2. "The road to hell has a silver lining."
3. "The worm is on the other foot."
4. "In one ear and gone tomorrow."
5. "A bird in the hand waits for no man."

17. BY ANOTHER NAME

"A rose by any other name would smell as sweet," wrote Shakespeare in *Romeo and Juliet*.

_____ by any other name would smell as _____.

Can you spoof this proverb by filling in the above blanks with new words that *rhyme* with "rose" and "sweet," respectively?

18. Spaced Out

A whimsical astronomer amused himself by writing the following sequence of words: Thermometer, Armless, Dirt, War, Lightning, Rings, Posterior, Sea. He added the word "Underworld," but then crossed it out. What does this list mean?

19. Opposites

Fill this baby crossword grid with words that are the *opposite* of the definitions given. If you do it correctly, the result will be a word square with the same words vertically as horizontally.

1	2	3	4
2			
3			
4			

1. Shrink
2. Green
3. Shut
4. Came

20. Make It Right

In this sentence there are neither more nor less than _____ three-letter words. What can you put in the blank space to make the sentence a true one? (You can't put *two*, for example, because that would give the sentence *three* three-letter words.)

21. An Odd Fish

A spelling reformer of the 19th century, appalled at the irregularities of English spelling, once commented that the word *fish* could be spelled *ghoti*. Can you deduce his rationale for saying that *ghoti* is a logical alternate spelling for *fish*?

22. Find the Word

Circle a letter in each of the above squares so that the circled letters, left to right, spell a common word.

23. Mystery Hieroglyphics

What do these strange symbols mean?

24. ENINYOS

A photograph shows a gaggle of little children at a playground. Add the same letter to four different places in the letter sequence ENINYOS to make an appropriate caption for the picture.

25. What Letter?

The three black shapes shown above represent a letter. What letter?

IMPONDERABLES

How can there be self-help *groups?*

Why are "wise man" and "wise guy" opposites?

Is it possible to be totally partial?

What's another word for *thesaurus?*

What was the greatest thing *before* sliced bread?

What does it mean to "pre-board?" Do you get on before you get on?

What is a "free" gift? Aren't all gifts free?

If lawyers are disbarred and ministers defrocked, are electricians delighted, musicians denoted, cowboys deranged, models deposed, tree surgeons debarked, and dry cleaners depressed?

26. HIDDEN WISDOM

In each sentence below a word is concealed, such as *no* in the fifth sentence. If you can spot the other buried words they will spell a well-known proverb when taken in order.

1. The word buried here has only one letter.
2. He looks like the troll in Grimm's fairy tales.
3. They're the best ones I've ever seen.
4. Are you looking at her shoes or her legs?
5. He's an old friend.
6. My thermos spilled coffee when I opened it.

27. Windbag's Gift

Professor Windbag likes to use big words and say everything in the most complicated way he can. When he handed a birthday present to his wife he said: "My dear, here is a diminutive, aurum, truncated cone, convex on its summit and semi-perforated with symmetrical indentations and a hollow interior." Can you guess what is in the box?

28. Jury Members

Three men on a jury panel were named Brown, Black, and Blue. One wore brown socks, one black socks, and one blue socks.

"Isn't it odd," said the man with black socks, "that none of us wears socks with a color that matches his name?"

"By golly, you're right!" exclaimed Mr. Brown.

What color of socks was each man wearing?

29. Tricky Triangle

See if you can correctly read aloud the words inside the above triangle. Almost everyone asked to do this seemingly simple task makes a mistake.

30. A Remarkable Fellow

In a "Dr. Matrix" column for *Scientific American*, I once introduced a character named Jasper Whitcomb Lundy. Can you find two unusual things about this gentleman's name?

Starting at one of the letters above, and then moving like a chess king in any of the eight directions to an adjacent square, you'll find you can spell every number from *one* through *nine*! (You may stand on the same letter twice to double it.) After testing this statement, see if you can spell:

1. The last name of a president of the United States.
2. The last name of a prime minister of India.
3. The name of an African country.
4. The pen name of an American short-story writer.

ILLITERACY

Do you read Poe?
No, I read pretty good.

Do you like Kipling?
Don't know. I've never kippled.

Have you read Marx?
No, not since I had the measles as a kid.

---

32. A New Recipe

In his little book *Puzzling Posers* (London, 1952), J. Travers gives the 5×5 letter square reproduced below. The puzzle is to imagine a chess king placed on any letter and moved to spell out a familiar motto, using each letter exactly once.

Can you find the popular saying? And if you allow reusing letters, as well as standing on a letter twice in a row to double it, can you construct a 4×4 letter square that contains the same saying?

33. A Secret Greeting

Can you decode the message from me to you that's concealed in this picture?

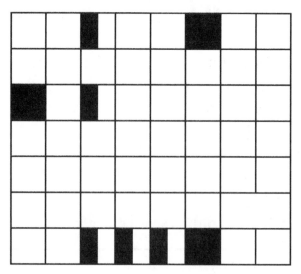

34. A SHORT STATE-MENT

How well do you know the two-letter postal abbreviations for the U.S. states? Find the state whose abbreviation matches each definition on the list below. For example, the answer to the first one is Alabama, for AL.

1. A man's first name
2. Either/_____
3. Physician
4. Musical note (two answers)
5. Driver's license, for example
6. Greetings
7. Pa's wife
8. Egomaniac's concern
9. All right
10. Exclamation of surprise
11. Business firm
12. Chic

35. NATIONWIDE TOUR

Besides those post office abbreviations, many states have older, more informal abbreviations. A farmer may choose to live in Delaware, because its abbreviation is "Del." With that example in mind, tell which state these people belong in.

1. A hypochondriac
2. A swindler
3. Noah
4. A laundromat owner
5. Bo Derek
6. A Catholic
7. A miner
8. A young lady

36. WORD, WARD, WARE, DARE, DAME, GAME

On Christmas day, 1877, Lewis Carroll invented a game he called Word Links to entertain two bored little girls. He later changed the name of his game to Doublets, and it became a parlor craze in London. Today the game is sometimes called "word ladders" or "word golf."

The object is to transform one word into another, related word by changing one letter at a time and producing a new word at each step. For example, Carroll turned HEAD to TAIL in this way:

$$\begin{array}{cccc} H & E & A & D \\ H & E & A & L \\ T & E & A & L \\ T & E & L & L \\ T & A & L & L \\ T & A & I & L \end{array}$$

Here are two Carroll challenges reprinted from the articles on Doublets he wrote for *Vanity Fair* magazine in 1879. First, can you "drive PIG into STY"? Carroll's solution has five transformations. Then try to "raise FOUR to FIVE." Carroll did it in seven steps.

It has been observed that doublets resemble the way in which evolution creates a new species by making small random changes in the "genes" that are intervals along the helical DNA molecule. Carroll himself, though a skeptic of Darwin's theory, evolved MAN from APE in six steps. Can you do it in five?

37. Mischmasch

Carroll also invented a little-known word game he called Mischmasch. The first player starts with a combination of letters called a "nucleus"—*GP*, for example. The second player tries to find a common word that contains that nucleus. Answers for *GP* might include *magpie*, *bagpipe*, *flagpole*, or *gangplank*.

Here are some "nuclei" that Carroll sent his godson. Can you find words for all of them?

1. NGU
2. IMSE
3. MFI
4. EWH

38. Wrong Signs

This is an easy one. In each word of the following five signs a single letter has been changed:

| QUIST | | PUAH | | EPIT | | FOX SAME | | EET PLINT |

Correct each sign.

39. Unreadable

Believe it or not, each of these series of words is a perfectly sensible sentence if punctuated properly. Can you provide the right punctuation?

That that is is that that is not is not is that it it is

Ed where Bill had had had had had had had had had had had the editor's approval

40. SOMETHING MISSING

A	B	C	D	E	F	G	H
I	J	K	M	N	O	P	Q
R	S	T	U	V	X	Y	Z

Suggested by the above chart are:

1. A Christmas carol.
2. The place that inspired the title for Samuel Butler's famous novel *Erewhon.*

What's the greeting and the place?

41. ETAOIN SHRDLU

When the American linotype machine was invented, it was believed that the twelve most commonly used letters in English, in decreasing order of frequency, were *etaoin shrdlu.* The letters in this nonsense phrase were spelled by the first and second columns of the traditional linotype keyboard. Before newspapers went digital, a printer would sometimes run a finger down the columns to make a slug, for marking purposes, that he intended to remove later. If he forgot to remove it, the cabalistic words *etaoin shrdlu* would mystify readers by appearing in the next day's paper.

There is only one English word that is an anagram of the letters of *etaoin shrdlu.* It is found in the *New Standard Unabridged Dictionary,* where its meaning is given as "more bizarre." It is also possible to use the letters to make two words that name a region of a world nation. Can you discover the word and the region?

MORE LINGUISTIC CATCHES

Ask someone, "How do you pronounce P-O-S-T?" Then, "How do you pronounce R-O-A-S-T?" Finally, "What is it you put in a toaster?" If he says, "Toast," say, "That's odd. I put in bread!"

Ask someone to write any word on a piece of paper, fold the paper, then stand on it. Say: "Using my psychic powers I'll tell you what's on the paper." Concentrate a moment, then say "*You* are on the paper." Follow by stating, "And now I'll tell you where you got those shoes you are wearing." "Where?" "On your feet."

"How do you pronounce the capital of Kentucky— 'Loo-ey-ville' or 'Loo-iss-ville'?" Whatever the victim replies, you say, "Wrong! The correct pronunciation is 'Frankfort.'"

42. BLACK EYE

In each of the eight sentences below the same single word is missing, and moving it around in the sentence can dramatically change the meaning. What word is it?

_____ I hit him in the eye yesterday.
I _____ hit him in the eye yesterday.
I hit _____ him in the eye yesterday.
I hit him _____ in the eye yesterday.
I hit him in _____ the eye yesterday.
I hit him in the _____ eye yesterday.
I hit him in the eye _____ yesterday.
I hit him in the eye yesterday _____.

43. Career Choices

On the left of the list below are the first names of 16 women. On the right are 16 professions. Each woman belongs to one of the professions. For example, Sue is an attorney, fittingly enough. See if you can match each name on the left with a related profession on the right.

Sue	Chiropractor
Grace	Waitress
Bridget	Upholsterer
Patience	Engineer
Carlotta	Dancer
Robin	Thief
Ophelia	Physician
Wanda	Milliner
Sophie	Minister
Hattie	Singer
Octavia	Magician
Carrie	Gambler
Betty	Musician
Carol	Automobile salesperson
Faith	Jeweler
Pearl	Lawyer

44. Three Little Letters

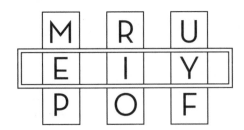

Mentally slide the three strips up and down until a familiar three-letter word appears inside the window.

45. Smith's Plans

"What are your plans for this afternoon?" asked Mrs. Smith.
"I intend to wash and polish our car," Smith replied.
Can you do the following?

1. Add a letter to a one-syllable word in the above dialog that will turn it into a three-syllable word.
2. Capitalize the first letter of a word to change its pronunciation *and* meaning.
3. Rearrange the letters of *intend* to make three different six-letter words.

46. Toothpick Teaser

Arrange 20 toothpicks (or matches) like so:

Remove nine toothpicks (without rearranging) to spell the name of a president.

47. Orbital Order

If you list the first letters of the eight planets in order of their distances from the sun, the sequence contains two surprising coincidences. Can you find:

1. Letters (out of order) that spell the name of a planet?
2. Letters (in order) that spell the name of another member of our solar system?

48. Go Jump in a Lake!

A sign by a lake said, "Private. No Swimming Allowed." A group of children who wanted to go swimming in the lake cleverly added punctuation to the sign to make it appear as if it permitted them to swim. How did they do this?

49. Colors and Geography

By starting at any square in the above grid, then moving one square in any direction (like a chess king), see if you can find the names of six colors. (Standing on a letter to double it is permitted, as is reusing letters.) Then see if you can spell:

1. A nation in the British Isles
2. A European capital city
3. A country in north Africa
4. A region in India and Bangladesh
5. A nation on the Persian Gulf
6. A country in the Balkans

I was unable to fit all the letters needed for spelling the six colors inside an order 4 matrix without duplicating one letter. If any reader can do this or prove it impossible, please send the pattern or the proof.

50. Penny for Your Thoughts

Many words are suggested or concealed on a penny. Carefully study each side of a penny and see if you can locate an indication in some way of names for the following:

1. A car
2. A snake
3. A mammal
4. A fruit
5. An item of furniture
6. A vehicle
7. A heavenly body
8. A bone of your body
9. A girl's first name
10. Five boys' first names

51. The Rectangular Pig

With paper matches you can make a picture of a pig as shown above. (Use a match head for the eye, and curl the tail for effect, if you like.) The puzzle is this: Change the positions of just two matches and the eye to make the rectangular pig look the other direction.

52. OPEN THE SAFE

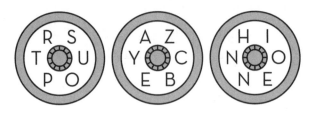

You see here the three dials of a locked safe. To open the safe you must rotate the dials until a familiar six-letter word can be read horizontally like the letters TUYCNO in the illustration. (Some or all of the letters won't be right-side-up once the dials are turned.)

53. NOTABLES

Imagine a chess king placed on the mini-board above. By moving the king one step in any direction (horizontally, vertically, or diagonally), you can spell the full name of a U.S. president. (You may reuse letters, or stand on a square two "moves" in a row to make a double letter.)

Start the king on other cells and see if you can spell:

1. The last names of two other U.S. presidents.
2. The first name of a British prime minister.
3. The first name of an American novelist, which is also the last name of a *different* American novelist.
4. The last names of two prizefighters, who first fought in 1964.
5. The last name of a female novelist born in 1832.

54. Mrs. Ford's Sculpture

These four pieces of metal sculpture, mounted on a wooden base, are the work of Mrs. Elinor Louise Ford, an abstract artist. Her work is saying something, but what?

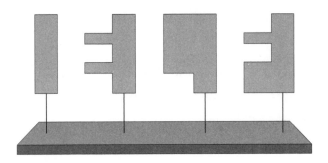

55. Add the O's

In each of the following meaningless groups of letters, can you insert three or four O's to make a common word?

1. VD
2. SNRUS
3. FFSHT
4. CKBK
5. PRTCL
6. LKUT
7. MNTNUS
8. STRERM
9. RATRI
10. CRRBRATR
11. SCILGY
12. DUBLN
13. BH
14. DRUS
15. RCC
16. FRENN
17. LNG
18. FTLSE
19. RTHDX
20. HMLGUS

56. Hix Nix Stix Pix

The cryptic sentence above is the title of a 1984 novel by David Burdett, who adapted it from a famous 1935 headline in the Hollywood trade paper *Variety*. Can you deduce the upshot of the newspaper story so mysteriously headlined?

WALKING INTO A BAR

A man walks into a bar holding a slab of asphalt and says, "A beer please, and one for the road."

A toothless termite walks into a bar and asks "Is the bar tender here?"

A neutron walks into a bar and orders a beer. The bartender says, "For you, no charge!"

Charles Dickens walks into a bar and orders a martini. The bartender asks, "Olive or twist?"

A jumper cable walks into a bar. The bartender says, "I'll serve you, but don't start anything."

Shakespeare walks into a bar and orders a beer. "Can't serve you," says the bartender. "You're still Bard here."

Two peanuts walk into a bar. One of them was a salted.

A hamburger walks into a bar and the bartender says, "Sorry, we don't serve food here."

A hydrogen atom walks into a bar and says, "I lost my electron in here." The bartender asks, "Are you sure?" "Yes," the atom replies, "I'm positive!"

A grasshopper walks into a bar. The bartender says, "Wow, funny you should come in here. We have a drink named after you." The grasshopper says, "You have a drink called Stanley?"

A priest, a preacher, and a rabbi walk into a bar, and the bartender says, "What is this, a joke?"

(Thanks to *Gilbert Magazine* for most of the above.)

57. Split Up the Family

The names of four family members are in the above rectangle. It's easy to draw three straight lines that will put each name in a separate compartment:

Do the same thing with just *two* straight lines.

58. Three of a Kind

For each triplet of words listed below, think of a fourth word that can be paired with all three to make a compound word or familiar phrase. For example, the answer for *boat, shelf, sentence* would be *life* (lifeboat, shelf life, life sentence).

1. Surprise, line, search
2. Base, snow, room
3. Cake, blue, cottage
4. Nap, wild, call
5. Golf, foot, night
6. Tiger, news, weight
7. Painting, lady, nail
8. Blood, beet, loaf
9. Show, ocean, plan
10. Light, village, thumb

59. Air Mail

The cryptic message below appeared on a postcard from a friend who has lived overseas for years. What's the hidden meaning?

A, B, ...

....

....

60. Straight Arrows

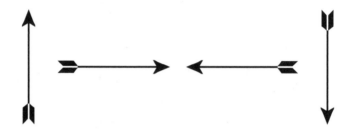

The symbols above stand for a familiar four-letter word. What word?

61. O, I C

Things can be named by a single letter. For example, *K* might be a girl's name (Kay). What letters name the following?

1. A part of the body
2. An insect
3. A drink
4. A vehicle
5. A piece of sports equipment
6. A vegetable
7. An animal
8. A body of water
9. An exclamation
10. A bird

RHETORICAL QUESTIONS

Does the name Pavlov ring a bell?

Would a book about voyeurism be a peeping tome?

If a pig has laryngitis, does he get disgruntled?

Is Karl Marx's tomb a communist plot?

I had amnesia once—or was it twice?

62. FROM A TO Z

On what basis are the letters of the alphabet partitioned, as shown in the following examples?

1.

A				E	F		H	I		K	L	M	N						T		V	W	X	Y	Z
	B	C	D			G			J					O	P	Q	R	S		U					

2.

A	B		D											O	P	Q	R								
		C		E	F	G	H	I	J	K	L	M	N					S	T	U	V	W	X	Y	Z

3.

		C				G		I	J		L	M	N					S		U	V	W			Z
A	B		D	E	F		H			K				O	P	Q	R		T				X	Y	

4.

A					F		H	I	J	K	L	M	N	O		Q	R	S		U		W	X	Y	
	B	C	D	E		G									P				T		V				Z

63. FIVE EASY PIECES

Here's a puzzle to try in person. Ask a friend to take away 5 toothpicks from this arrangement and leave 8. (Or try it yourself, but be careful!)

64. THE BREAKUP

<u>UALLS</u>
NOW

What sentence is represented by the above rebus?

65. FOUR-LETTER WORDS

A _____ old woman on _____ bent
Put on her _____ and away she went.
"Come, _____, my son," she was heard to say.
"What shall we _____ upon today?"

Fill in each blank with a four-letter word, each word using the same four letters. Can you do the same with the blanks in the following adaptation of an old English drinking song?

Come, landlord, fill the flowing _____
Until the _____ run over,
We'll _____ and drink it on this _____.
And then we'll _____ to Dover.

Is there a sixth word that uses the same four letters?

66. Mind Games

B _ R _ A _ I _ N

_ B _ R _ A _ I _ N

Put one letter in each of the blanks to make two words.

67. Square Root

Palindromist Leigh Mercer, whom we met in the first chapter, once sent me this list of four passable sentences. What unusual characteristic do these sentences share?

> Just ugly slip type.
> Might idler glide hedge trees?
> Crest, reach eager (scene three).
> Leave Ellen alone, venom enemy!

68. Gender Confusion

During a brief visit to America, a man who could read no English met a friend for lunch. The foreigner excused himself to use the restroom, but paused to ask his friend, "How can I tell which is for men and which for women?" "Easy," the friend replied. "Just go into the room that has the shorter name on the door." But this seemingly foolproof advice backfired. Can you guess why?

69. Be Back Soon

When his shop was open, an antique dealer would put in his front window four large letter cards that spelled OPEN. But the dealer had no CLOSED sign. What simple task would he perform to unmistakably show customers that his shop *wasn't* open?

70. A Very ALNSUUU Dictionary

In my *Scientific American* column of November 1965, I printed Nicholas Temperley's suggestion for an unusual dictionary: one that would list each word with its letters in alphabetical order. Who would consult such a dictionary? Anagram seekers, of course! The entry for AEGINLRT, to take one example, would be followed by *integral*, *relating*, and *triangle*. In the decades since my column first appeared, several publishers, including Follett and Longman, have actually produced complete anagram dictionaries following this principle.

In a dictionary of this kind, the first entry would be A, for *a*, followed by AA for *aa*, a type of lava. Assuming the dictionary only contains common English words, you might be surprised to find that the third word in the dictionary has eleven letters! What is it?

Temperley estimated that the *A* section of his dictionary, unlike that of a conventional one, would make up half of the volume. The sections for *B*, *C*, and so on would be progressively shorter, and the lengths drop off precipitously after *I*. What two articles of clothing are most likely the last two words in the dictionary?

71. Ten Commandments, Thirty Consonants

This cryptic inscription is said to have appeared over the Ten Commandments in a medieval church:

PRSVRYPRFCTMN

VRKPTHSPRCPTSTN

Can you decipher the sculptor's message by adding just one letter?

72. Two by Two

What pair of letters comes next in this series?

ST, ND, RD, TH, ...?

73. Neighbors

There is only one set of four consecutive letters of the alphabet that can be anagrammed into a common four-letter word. What is it?

There are two such set of three consecutive letters, and finding them is a much easier problem.

74. Postcard from Paraguay

A postcard arrived in the mail from Asunción, the capital of Paraguay, so mangled from its journey from South America that only the last four letters of its final word were legible: -*cion*. Besides Spanish words like *Asunción*, can you think of three common *English* words that end with the letters -*cion*?

75. Give Yourself a Hand

You are solving a crossword, and see a clue that reads "Number of fingers." You look at the grid, wondering if *five* or *ten* will be the word that fits. To your surprise, the answer seems to be nine letters long, and starts with a *J*! What do you write in the blanks?

Answers to Chapter 2

1.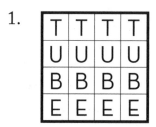

2. Incomprehensibilities.

3. The "blind man" owns a shop that sells window blinds.

4. Oyster, turtle.

5. Cover the top halves of the symbols with a card or paper sheet. You'll see the word *JULY*.

6. *Stop* and *snap* backward spell *pots* and *pans*. Nitram Rendrag in reverse is *Martin Gardner*.

7. The names are all anagrams of U.S. states: New York, North Carolina, California, North Dakota, Delaware, Rhode Island, Nebraska, New Jersey.

8. Exist, blank, center.

9. Shift each letter of *HAL* forward one step in the alphabet and you get *IBM*. Clarke insisted this was sheer coincidence, not something he intended.

10. She saw ten, since you can shift the letters in *UFO* one step backward in the alphabet to spell *TEN*.

11. The words are *drive* and *park* (or vice versa).

12. One possibility is this quote from a child before bed: "What did you bring that book I didn't want to be read to out

of up for?" If the bedtime story is about Australia, the string of prepositions can be even longer: "What did you bring that book I didn't want to be read to out of about Down Under up for?"

13. Queue (and the less familiar word *aitch* works as well)

14. Nth. The longest common words I could find using the odd letters of the alphabet are ten-letter ones like *acquiesces* and *cockamamie*; long odd-letter phrases include *Mickey Mouse* and *easy come, easy go*.

15. William (note the double letter in his name) likes things with double letters in their names.

16. (1) The early bird saves nine. (2) Every cloud is paved with good intentions. (3) The shoe turns. (4) Here today, out the other. (5) Time and tide is worth two in the bush.

17. Toes, feet

18. The words suggest the eight planets in their order from the sun: Mercury, Venus, Earth, Mars, Jupiter, Saturn, Uranus, and Neptune (with Pluto crossed out).

19.

G	R	O	W
R	I	P	E
O	P	E	N
W	E	N	T

(Thanks to Tim Tebbe.)

20. Put the numeral 2 in the blank.

21. Pronounce the *gh* as in *rough*, the *o* as in *women*, and the *ti* as in *nation*, and the resulting combination sounds like *fish*. Similar logic can be used to show that *ghghgh* should be pronounced *puff*: the first *gh* as in *hiccough*, the second *gh* as in *Edinburgh*, and the third *gh* as in *laugh*.

22. FLY.

23. On the left side of each symbol are the initials of the days of the week. On the right side are their mirror reflections.

24. Add four *T*'s to make TEN TINY TOTS.

25. Give the page a quarter turn counterclockwise. You'll see the letter E.

26. A rolling stone gathers no moss.

27. A gold thimble.

28. Brown had blue socks, Blue had black socks, and Black had brown socks. Proof: Brown can't have brown socks because they would match his name. He can't have black socks because they are on the man who asked the question that Brown answered. Therefore Brown's socks must be blue. This leaves black socks for Mr. Blue and brown socks for Mr. Black.

29. "Paris in the the spring." Did you miss the repeated *the?*

30. His name contains 19 of the 26 letters of the alphabet, without a single repetition. It also includes the six vowels in alphabetical order.

31. (1) Eisenhower, (2) Nehru, (3) Niger, (4) O. Henry.

32. The saying is "Too many cooks spoil the broth." Here is one possible 4×4 square that contains the same proverb (allowing repeated letters by standing on a letter):

L	T	K	S
H	I	O	P
E	R	M	C
B	A	N	Y

33. Hold the book flat and view the picture on a strong slant. You'll see the greeting "Hello!" (Thanks to Marvin Miller.)

34. (1) Alabama (AL), (2) Oregon (OR), (3) Maryland (MD), (4) Louisiana (LA) or Michigan (MI), (5) Idaho (ID), (6) Hawaii (HI), (7) Massachusetts (MA), (8) Maine (ME), (9) Oklahoma (OK), (10) Ohio (OH), (11) Colorado (CO), (12) Indiana (IN)

35. (1) Illinois (Ill.), (2) Connecticut (Conn.), (3) Arkansas (Ark.), (4) Washington (Wash.) (5) Tennessee (Tenn.), (6) Massachusetts (Mass.) (7) Oregon (Ore.), (8) Mississippi (Miss.)

36. Carroll's solutions were PIG-BIG-BAG-BAY-SAY-STY and FOUR-FOUL-FOAL-FOAM-FORM-FIRM-FIRE-FIVE. Carroll's APE-MAN solution was APE-ARE-ERE-ERR-EAR-MAR-MAN, but a shorter chain is APE-APT-OPT-OAT-MAT-MAN.

37. (1) Tongue, (2) Himself, (3) Discomfited, (4) Somewhere. Other answers are possible.

38. Quiet, Push, Exit, For Sale, Wet Paint.

39. That that is, is. That that is not, is not. Is that it? It is.

Ed, where Bill had had "had," had had "had had." "Had had" had had the editor's approval.

40. (1) Noel ("no L"), (2) Nowhere ("no W here").

41. Outlandisher, South Ireland.

42. Only.

43. Sue — lawyer

Grace — dancer

Bridget — engineer

Patience — physician

Carlotta — automobile salesperson

Robin — thief

Ophelia — chiropractor

Wanda — magician

Sophie — upholsterer

Hattie — milliner

Octavia — musician

Carrie — waitress

Betty — gambler

Carol — singer

Faith — minister

Pearl — jeweler

44. PRY.

45. (1) Area, (2) Polish, (3) Indent, dentin, tinned.

46.

47. The initials MVEMJSUN (Mercury, Venus, Earth, Mars, Jupiter, Saturn, Uranus, and Neptune) contain the letters of (1) Venus (nonconsecutively), and (2) the Sun (consecutively).

48. "Private? No! Swimming Allowed."

49. The colors are the colors of the rainbow: red, orange, yellow, green, blue, purple. (1) England, (2) Riga (Latvia), (3) Algeria, (4) Bengal, (5) Iran, (6) Bulgaria.

50. (1) Lincoln, (2) Copperhead, (3) Hare (Lincoln's "hair," homophonically) (4) Date (the year), (5) Sofa, in "United States *of A*merica," (6) A bus in "E Pluri*bus* Unum," (7) Sun, in "E Pluribu*s Un*um," (8) Rib, in "E Plu*rib*us Unum," (9) Erica, in "United States of Am*erica*," (10) Bert in "li*bert*y," Ted in "Uni*ted*," Ed in "Unit*ed*," Eric in "Am*eric*a," and Tate in "S*tate*s." (You might also have noticed less common names like Ingo in "*In Go*d We Trust," Uri in "E Pl*uri*bus Unum," and Ty in "Liber*ty*.")

51.

52. PSYCHE.

53. William Jefferson Clinton is the president. You can also spell (1) Lincoln (and Jefferson, of course), (2) Winston (Churchill), (3) Sinclair (Sinclair Lewis and Upton Sinclair) (4) (Muhammad) Ali and (Sonny) Liston, and (5) (Louisa May) Alcott.

54. Place a sheet of dark paper above and below the dark shapes. You'll see the word *ELF*, Mrs. Ford's initials.

55. (1) Voodoo, (2) sonorous, (3) offshoot, (4) cookbook, (5) protocol, (6) lookout, (7) monotonous, (8) storeroom, (9) oratorio, (10) corroborator, (11), sociology, (12) doubloon, (13) boohoo, (14) odorous, (15) rococo, (16) forenoon, (17) oolong, (18) footloose, (19) orthodox, (20) homologous.

56. The headline (originally "Sticks nix hick pix" in *Variety*) meant that rural moviegoers were, contrary to conventional Hollywood wisdom, rejecting movies about rural life.

57.

F A T	H E R		
H E R	M A N		
L O U	I S A		
I S A	B E L		

58. (1) Party, (2) Ball, (3) Cheese, (4) Cat, (5) Club, (6) Paper, (7) Finger, (8) Sugar, (9) Floor, (10) Green.

59. "Long time, no see" (no *C*)

60. News. (Abbreviations for the four points of the compass: north, east, west, south.)

61. (1) I, (2) B, (3) T, (4) L, (5) Q, (6) P, (7) U, (8) C, (9) G or O, (10) J.

62. (1) The top letters all have straight lines, the lower letters have curved lines. (2) The top letters have holes. (3) If the top letters were made of elastic, each could be stretched to a straight line. (4) The bottom letters all rhyme with *E*.

63.

H T E

64. All between us is over now.

65. Vile, evil, veil, Levi, live. Pots, tops, stop, spot, post. The sixth word is *opts*.

66. *Barbarian* and *aberration*. (Thanks to Chris Masianisa.)

67. The words in each sentence can be placed in a word square so that the same sentence is read down the columns as across the rows.

68. Unfortunately, the two doors read *Ladies* and *Gentlemen*.

69. He would move the final card around to the start of the word, so his window read NOPE.

70. The third entry is AAAAABBCDRR, for *abracadabra*, unless the dictionary compiler is a lover of Greek food and includes AAAAAALMRSTT for *taramasalata*. The last entries are quite probably TTUU and TUX, for *tutu* and *tux*.

71. If the letter *E* is added SVNTN—I mean, seventeen—times, the message reads, "PERSEVERE YE PERFECT MEN: EVER KEEP THESE PRECEPTS TEN."

72. The next sixteen pairs in the series are also *TH*. These are the suffixes that turn cardinal numbers into ordinals: 1st, 2nd, 3rd, 4th, and so on.

73. RSTU can be anagrammed to *rust* or *ruts*. (Thanks to Murray W. Pearce.) The three-letter combinations are found at the very beginning of the alphabet: ABC for *cab* and DEF for *fed*.

74. Scion, suspicion, coercion.

75. The nine-letter answer is *Jack Frost*, who can certainly numb your fingers.

3
BIZARRE POEMS

Puzzlesmiths like to make things difficult for themselves. The ordinary constraints of language—grammar and meaning—are not tough enough, so they invent special rules: a grid that is filled with intersecting words, a sentence that reads the same backward and forward, a paragraph that is written using only the words in the Pledge of Allegiance.

A poet has similar instincts. Though his primary concern is refining sound and sense, he makes his job harder by self-imposing schemes of rhyme, meter, or alliteration.

That a poet should turn puzzler, or vice versa, is a natural transformation—and one with a long history. Almost 2,500 years ago, the Greek poet Pindar wrote an ode without using the letter sigma; another Greek poet, Tryphiodorus, composed a 24-volume epic about Ulysses, each book omitting one letter of the Greek alphabet.

Hundreds of years later, in 15th-century Persia, the renowned poet Jami was approached by a lesser poet who wanted to read the great man a rhyme he had written.

"This work is quite unusual," the lesser poet proudly stated when he was done reading. "The letter *alif* is not to be found in any of the words!"

"You can do better yet by removing *all* the letters," was Jami's curt rejoinder.

SEVEN PUZZLE POEMS

The seven poems here each represent a particular type of wordplay. Can you determine what is remarkable about the structure of each poem?

1. Square Poem

I often wondered when I cursed,
Often feared where I would be—
Wondered where she'd yield her love,
When I yield, so will she.
I would her will be pitied!
Cursed be love! She pitied me ...

—Lewis Carroll

2. I Will Arise

I
will
arise
and
go
now,
and
go—any damned place
   just to get away from
      THAT
     chair
     covered
     with
     CAT
     hair
—William Jay Smith

3. Capacity

Capacity 26 Passengers
  —*sign in a bus*

Affable, bibulous,
corpulent, dull,
eager-to-find-a-seat,
formidable,
garrulous, humorous,
icy, jejune,
knockabout, laden-
with-luggage (maroon),
mild-mannered, narrow-necked,
oval-eyed, pert,
querulous, rakish,
seductive, tart, vert-
iginous, willowy,
xanthic (or yellow),
young, zebuesque are my
passengers fellow.

                    —John Updike

4. Curious Acrostic

Perhaps the solvers are inclined to hiss,
Curling their nose up at a con like this.
Like some much abler posers I would try
A rare, uncommon puzzle to supply.
A curious acrostic here you see
Rough hewn and inartistic tho' it be;
Still it is well to have it understood,
I could not make it plainer, if I would.

                    —Anonymous ("Maude")

5. *Winter Reigns*

Shimmering, gleaming, glistening glow—
Winter reigns, splendiferous snow!
Won't this sight, this stainless scene,
Endlessly yield days supreme?

Eyeing ground, deep piled, delights
Skiers scaling garish heights.
Still like eagles soaring, glide
Eager racers; show-offs slide.

Ecstatic children, noses scarved—
Dancing gnomes, seem magic carved—
Doing graceful leaps. Snowballs,
Swishing globules, sail low walls.

Surely year-end's special lure
Eases sorrow we endure,
Every year renews shared dream,
Memories sweet, that timeless stream.

—Mary Youngquist

6. *Night's Poem*

Idling, I sit in this mild twilight dim,
Whilst birds, in wild, swift vigils, circling skim.
Light winds in sighing sink, till, rising bright,
Night's Virgin Pilgrim swims in vivid light!

—Harry Leroy Temple

7. Spa

Laughing boys, legs bare, with girls bathing—
Girls kind of fond are these,
Chaffing and cheering boys, limbs writhing ...
Swirls water, whips spumes, splash seas,
   Breaking, into shrieking
      Girls ...
      Noise and boys,
      Boys and noise ...
      Girls,
   Shrieking, into breaking
Seas splash, spume whips, water swirls ...
Writhing limbs, boys cheering and chaffing—
These are fond of kind girls,
Bathing girls with bare legs, boys laughing.
   —J.A. Lindon

DID YOU HEAR THE ONE ABOUT ...

The dentist and manicurist who fought tooth and nail?

The seismologist whose theories turned out to be shaky?

The linguist who spoke Esperanto like a native?

The woman who found her first gray hair and thought she'd dye?

The dermatologist who built his business from scratch?

The man who worked in a tent factory, but it folded?

The corduroy pillows that are always making headlines?

The man whose feet are so big he has to put his pants on over his head?

Magic Verse

A strange collection of poems entitled *Shadows in the Moonlight* was written and privately printed in Los Angeles in 1927 by T. Page Wright, a Hollywood scriptwriter. Wright was a skillful amateur magician who wrote the poems so that the book could be used for performing a feat of mental magic. Each of the 22 poems is so constructed that the 19th word is "rose" and the 31st word is "love." The book originally sold in magic stores for five dollars, but is now extremely scarce. Considering the purpose for which the poems were written, they are surprisingly good. I particularly like this one:

At a Melodrama

The leading lady suffered, while I yawned;
But a girl beside
Me sat and cried—
Her tears that rose unbidden washed away
The time worn tawdriness that stained the play.
Love was the theme—the rest she looked
  beyond.

I constructed the following two poems for use in two magic prediction tricks in my Sterling book *Mental Magic* (1999). In the first poem the *n*th word of every *n*th line is "red." The prediction method for the second poem involves rolling a die. The number on top of the die indicates a line of the poem, and the number on the bottom of the die counts the words in that line. Regardless of the outcome of each die toss, the word counted to is always *blue*. Try it out.

In Praise of Red

Red is the color of sunsets and fire,
And red is our blood when it flows.

A beautiful red are the lips of my love.
They rival the red of a rose.

We thrill to the red of a cardinal's wings,
But not to a sunburned red nose!

In Praise of Blue

Barbara's eyes are a beautiful blue.
On a bluebell, a blue butterfly.
I seldom am blue in December.
Though frequently blue in July.
A blue cheese is tasty on salads.
Blue berries are great in a pie.
But the most wonderful blue of all
Is the blue of a cloudless sky.

THE BOX WITHOUT A LID

Lewis Carroll is the author of the following riddle:

John gave his brother James a box:
About it there were many locks.

James woke and said it gave him pain;
So he gave it back to John again.

The box was not with lid supplied,
Yet caused two lids to open wide:

And all these locks had never a key—
What kind of box, then, could it be?

Carroll provided the answer to this riddle-poem with another poem:

> As curlyheaded Jemmy was sleeping in bed,
> His brother John gave him a blow on the head;
> James opened his eyelids, and spying his
>   brother,
> Doubled his fist, and gave him another.
> This kind of box, then, is not so rare;
> The lids are the eyelids, the locks are the hair,
> And so every schoolboy can tell to his cost,
> The key to the tangles is constantly lost.

THREE LITTLE PUNS

The prefatory poem of Carroll's *Alice in Wonderland* opens with these lines:

> All in the golden afternoon
> Full leisurely we glide;
> For both our oars, with little skill,
> By little arms are plied,
> While little hands make vain pretence
> Our wanderings to guide.

Most of the many puns in the two *Alice* books are easy to spot, but the three puns in the above passage are often overlooked. The real Alice's last name was Liddell, rhyming with *fiddle*. Carroll is recalling the day when he took Alice and her two sisters on a rowing trip up the Isis River. All three occurrences of the word *little* in the passage are puns on *Liddell*.

MORE RIDICULOUS RIDDLES

1. How do you make a hippopotamus float?

2. What's red, then purple, then red, then purple …?

3. What weighs six tons and sings calypso?

4. What do you sit on, sleep on, and brush your teeth with?

5. What has eighteen legs, is covered with red spots, and catches flies?

6. What is round and green, is covered with blue hair, has big scaly claws, weighs five thousand pounds, and goes peckity-peck-peck?

7. What do you get when you cross an agnostic with a dyslexic?

8. Where do dictators keep their armies?

9. Why do cannibals not like to eat clowns?

10. How do you top a car?

ACROSTICS

Lewis Carroll was also a devotee of the acrostic, the art of concealing hidden messages using the first letters of lines of poetry. He wrote a raft of acrostic poems, for example, on the names of his female child friends, some of them providing dedicatory verses to a book. Here is his unusual acrostic dedication to Isa Bowman in the first volume of *Sylvie and Bruno:*

Is all our Life, then, but a dream
Seen faintly in the golden gleam
Athwart Time's dark resistless stream?
Bowed to the earth with bitter woe,
Or laughing at some raree-show,
We flutter idly to and fro.
Man's little Day in haste we spend,
And, from the merry noontide, send
No glance to meet the silent end.

Observe that the acrostic takes two different forms—the first letters of each line spell *Isa Bowman*, and the first three letters of the first line of each stanza are *Isa*, *Bow*, and *Man*. Carroll records in his *Diary* that Isa was not aware of the second acrostic until he told her about it.

Carroll was intrigued by acrostics before he started including them in his novels. In 1862, he had sent the following letter to Annie Rogers:

I send you

A picture, which I hope will
B one that you will like to
C. If your Mamma should
D sire one like it, I could
E sily get her one

Your affectionate friend
C.L. Dodgson

The earliest crude acrostics are found in the Old Testament, where nine Psalms are "abecedarian acrostics," the initial letters of each stanza consisting of the Hebrew letters in alphabetical order. If the 119th Psalm is read in the original

Hebrew, for example, each of the first stanza's eight verses begins with the letter *aleph*. The eight verses of the second stanza begin with *beth*, the verses of the third stanza with *gimel*, and so on through the 22 letters of the old Hebrew alphabet to *tav*, the final letter.

It was the Greeks and the Romans who introduced the sentence acrostics. The prophetic verses of the Greek Sibyls—old ladies who spouted hexameters while in a state of pretended religious frenzy—were often acrostics. And the fish only became a popular symbol of Christ because of an acrostic: the five initials of the five Greek words for "Jesus Christ, the Son of God, the Savior" spell the Greek word *ichthys*, meaning *fish*.

Acrostics blossomed during the Renaissance, but fell into disrepute in England after the Elizabethan period. Writing of acrostics and anagrams, Joseph Addison wrote that "it is impossible to decide whether the inventor of the one or the other were the greater blockhead." Samuel Butler, in his "Character of a Small Poet," described the acrostic writer as "one who would lay the outside of his verses even, like a bricklayer, by a line of rhyme and acrostic, and fill the middle with rubbish."

The following passage is from Act III, Scene 1, of Shakespeare's *A Midsummer Night's Dream*, spoken by the fairy queen Titania to Bottom the Weaver:

> **O**ut of this wood do not desire to go.
> **T**hou shalt remain here, whether thou wilt or no.
> **I** am a spirit of no common rate,
> **T**he summer still doth tend upon my state;
> **An**d I do love thee; therefore, go with me.
> **I**'ll give thee fairies to attend on thee,
> **A**nd they shall fetch thee jewels from the deep.

The bold italicized letters on the left spell *O Titania*. A coincidence? Or did Shakespeare, who enjoyed wordplay, intend the acrostic?

I once called this Edgar Allan Poe sonnet the finest acrostic by an American poet. It hides his beloved's name in an unorthodox way. Can you discover it?

> "Seldom we find," says Solomon Don Dunce,
> "Half an idea in the profoundest sonnet.
> Through all the flimsy things we see at once
> As easily as through a Naples bonnet—
> Trash of all trash!—how *can* a lady don it?
> Yet heavier far than your Petrarchan stuff—
> Owl-downy nonsense that the faintest puff
> Twirls into trunk-paper the while you con it."
> And, veritably, Sol is right enough.
> The general tuckermanities are arrant
> Bubbles—ephemeral and so transparent—
> But *this* is, now —you may depend upon it—
> Stable, opaque, immortal—all by dint
> Of the dear names that lie concealed within't.

Another unusual acrostic is "To Those Overseas," by J.A. Lindon of Addlestone, England.

> A merry Christmas and a happy new year!
> Merry, merry carols you'll have sung us;
> Christmas remains Christmas even when you
>   are not here,
> And though afar and lonely, you're among us.
> A bond is there, a bond at times near broken.
> Happy be Christmas then, when happy, clear,
> New heart-warm links are forged, new ties betoken
> Year ripe with loving giving birth to year.

The first acrostic on each line is easy to see, but Lindon has ingeniously worked into his poem a second pattern that is much less obvious. Can you find both?

MORE CLASSIC CLASSROOM CATCHES

Ask someone to make a fist with her left hand, then put her right hand palm down on top of the fist, and say "Wing" three times. Pick up her right hand, bring it to your ear, and say "Hello!"

"What's the difference between a mailbox and a garbage can?" If you get the answer "I don't know," say, "Well, I'll never give *you* a letter to mail!"

"If you asked a postal clerk to put the stamp you just bought on an envelope, and he refused, what would you do?" If you get the response, "I'd put it on myself," say, "Wouldn't it do more good to put the stamp *on the envelope?*"

Among 20th-century American authors, James Branch Cabell was particularly fond of acrostics. The prefatory poem of his novel *Jurgen* is an acrostic spelling the name of "Burton Rascoe," a critic who had championed Cabell's work. "The Sonnet Made for Maya," the last poem of Cabell's privately printed *Sonnets from Antan*, 1929, is an acrostic that spells "This is nonsense."

Aleister Crowley, the mad occultist and poet who has been the subject of several recent lurid biographies, enjoyed composing pious Protestant hymns and sending them to unsuspecting clerics. Not until the song had been sung for several Sundays would it dawn on some horror-stricken member of the congregation that the initial letters spelled some shocking anatomical request.

The poet John Peale Bishop was one of Crowley's modern heirs, since "A Recollection," one of his best-known poems, contains an obscene acrostic quite at odds with its lyrical tone. According to poet Theodore Roethke, an editor once included "A Recollection" in an anthology, and had to recall the book's first printing when the hidden profanities were discovered.

PROSE, BY ANY OTHER NAME

Today John Peale Bishop is better known as a good friend of his Princeton contemporary F. Scott Fitzgerald. The character of Tom D'Invilliers in Fitzgerald's semiautobiographical first novel, *This Side of Paradise*, is based on Bishop.

Do you notice anything odd about this excerpt from Chapter Three of *This Side of Paradise*?

> When Vanity kissed Vanity, a hundred happy Junes ago, he pondered o'er her breathlessly, and, that all men might ever know, he rhymed her eyes with life and death:
> "Thru Time I'll save my love!" he said ... yet Beauty vanished with his breath, and, with her lovers, she was dead ...
> —Ever his wit and not her eyes, ever his art and not her hair:
> "Who'd learn a trick in rhyme, be wise and pause before his sonnet there" ... So all my words, however true, might sing you to a thousandth June, and no one ever *know* that you were Beauty for an afternoon.

Though formatted as prose, the excerpt is actually a poem in disguise. It could be laid out as four stanzas, each with an *abab* rhyme scheme.

The title page and introduction of James Russell Lowell's famous satire *A Fable for Critics* appear to be the only parts of the book written in prose—but are they? A closer look reveals that the text is actually made up of tightly rhymed anapests. Even the copyright notice is written in verse:

> Set forth in October, the 21st day, in the year '48.
> G.P. Putnam, Broadway.

Washington Irving concealed 22 lines of iambic pentameter in the first paragraph of the sixth book of his *History of New York*, written under the pseudonym Diedrich Knickerbocker. (The poem begins: "But now the war-drum rumbles from afar. . . .") James Branch Cabell used a similar device in his novel *Jurgen*, mentioned above for its acrostical preface. In the 14th chapter, the title character sings,

> "Lo, for I pray to thee, resistless love . . . that
> thou to-day make cry unto my love, to Phyllida
> whom I, poor Logreus, love so tenderly, not to
> deny me love! Asked why, say thou my food and
> drink is love, in days wherein I think and brood
> on love, and truly find naught good in aught
> save love, since Phyllida hath taught me how to
> love. . . .
> "If she avow such constant hate of love as
> would ignore my great and constant love, plead
> thou no more! With listless lore of love woo
> Death resistlessly, resistless Love, in place of her
> that saith such scorn of love as lends to Death
> the lure and grace I love."

This is a sonnet in disguise, but an eccentric sonnet: each line ends with the word *love*, and the complicated rhymes are actually internal to the lines. See how many you can spot.

ACCIDENTAL VERSE

Some of the best disguised poems were unintentional. The most famous English example is in William Whewell's *An Elementary Treatise on Mechanics* (1819). It came to light during a dinner in Whewell's honor at Cambridge, where Whewell was Master of Trinity College. Adam Sedgwick, a geologist, rose and asked if anyone knew who had written the following stanza:

> And hence no force, however great,
> Can stretch a cord, however fine,
> Into a horizontal line
> That shall be absolutely straight.

Although the same stanza form had been used by Tennyson in "In Memoriam," no one could identify those particular lines. Sedgwick then revealed that he had quoted from page 44 of the first volume of Whewell's treatise, taking the liberty to polish the last line, which originally read, "Which is accurately straight." Whewell was so unamused that he altered the passage to eliminate the verse in the next edition of the book. He later published a volume of poetry called *Sunday Thoughts and Other Verses*, but his accidental verse is the only one he wrote that is still remembered.

A splendid specimen of chance doggerel is found in Lincoln's second inaugural address:

Fondly do we hope,
Fervently do we pray,
That this mighty scourge of war
May speedily pass away.
Yet, if God wills
That it continue until ...

There are countless other examples of accidental verse in all major languages. Here are two I stumbled on myself in *The New York Times*. James Thurber ended his article "The Quality of Mirth" (February 21, 1960) as follows:

If they are right and we are wrong,
I shall return to the dignity
Of the printed page, where it may be
That I belong.

And the first paragraph of James Reston's column "Mr. Ford's Last Chance" (January 16, 1976) ends with

This is the sound of prominent men,
Prodded by their wives,
Cleaning out the attic
And fleeing for their lives.

Some bits of accidental verse were clearly destined for a broader audience. Future Supreme Court justice William O. Douglas was riding the New Haven Railroad in the early 1930s when he and his friend noticed the hidden poetic qualities of a sign hanging in the washrooms of a Pullman car:

> Passengers will please refrain
> From flushing toilets while the train
> Is standing in
> Or passing through
> A station.

Douglas and his friend set the lines to a melody from one of Dvořák's *Humoresques*, and a schoolyard classic was born.

Herman M. Frankel passed along this gem from Werner Heisenberg's *Physics and Philosophy*:

> Every word or concept
> Clear as it may seem to be,
> Has only a limited range
> Of applicability.

Arthur Koestler, in *Roots of Coincidence*, quotes a passage from another physicist that can be read as follows:

> Particles of
> Imaginary mass,
> Interacting together
> Like a frictionless gas.

My wife Charlotte spotted this final item in a 1986 AP dispatch about road conditions in Windsor Heights, West Virginia:

> Neither snow nor rain, nor heat,
> Nor gloom of night,
> Stops the mail from getting through,
> But potholes might.

GREAT PUTDOWNS

I once heard Alfred North Whitehead lecture at the University of Chicago. He is today best known for collaborating with Bertrand Russell on a pioneer work titled *Principia Mathematica*. Whitehead opened his speech by saying that he and Bertie had many disagreements. "He thinks I'm awfully simple, and I think he's simply awful," or maybe it was the other way around.

Years earlier, while participating in a philosophical symposium, Whitehead famously thanked a previous speaker for "leaving the darkness of your topic unobscured."

Dorothy Parker, reviewing a Broadway play, described Katharine Hepburn's performance as "running the gamut of emotions from A to B."

During her feud with fellow author Lillian Hellman, Mary McCarthy announced on a talk show that "Every word [Hellman] writes is a lie, including *and* and *the*."

Ambrose Bierce, who disliked his publisher, suggested that the man's epitaph should read, "Here lies Frank Pixley—as usual."

And Milton Berle borrowed so much of his material from other comedians that Walter Winchell dubbed him "the thief of bad gags."

PUNCTUATED POEMS

Puzzles in which a nonsense statement is made sensible by altering punctuation can be found in many old puzzle books. Here is a poem that seems to describe impossible things:

Though seldom from my yard I roam,
I saw some queer things here at home.
I saw wood floating in the air;
I saw a skylark, bigger than a bear;
I saw an elephant with arms and hands;
I saw a baby breaking iron bands;
I saw a blacksmith, weighing half a ton;
I saw a statue sing and laugh and run;
I saw a schoolboy nearly ten feet tall;
I saw an oak tree span Niagara fall;
I saw a rainbow, black and white and brown;
I saw a parasol walking through the town;
I saw a politician doing as he should;
I saw a good man—and I saw some wood.

The poem makes sense if the semicolons are shifted to the middle of each line:

... I saw wood; floating on air
I saw a skylark; bigger than a bear
I saw an elephant; ...

... and so on to the end.

This anonymous poem doesn't seem to rhyme or scan. Can you re-punctuate it so that it does?

A Lady and Two Children

There was an old lady and she
Was deaf as a post.
A boy and girl behind a tree
Were kissing.

CLERIHEWS

Clerihews are short, rhyming poems that provide humorous biographical sketches of their subjects. They were invented in 1891 by the teenaged E.C. Bentley, who went on to become a successful mystery novelist. "Clerihew" was Bentley's middle name.

These are some of the best clerihews written by Bentley.

Daniel Defoe
Lived a long time ago
He had nothing to do, so
He wrote Robinson Crusoe.

George the Third
Ought never to have occurred.
One can only wonder
At so grotesque a blunder.

When Alexander Pope
Accidentally trod on the soap,
And came down on the back of his head—
Never mind what he said.

Bentley's classmate G.K. Chesterton (also a future novelist) was one of the first to try his hand at some clerihews of his own. Here are two modern clerihews I've borrowed from *Gilbert Magazine*, devoted to Chesterton and his work.

Blitzer's Fame

After Wolf Blitzer
Drank his eighth brandy spritzer,
He confessed that his fame
Was due entirely to his name.
—Jim Waters

Sandburg's Peril

Carl Sandburg
Sat on a iceberg
Miserable the entire time
Because he was so dangerously close to a
    rhyme.
> —Dale Alquist

A LOONY LEXICON

Here are five very long words. Can you show how to use each in a sentence by matching them to the blanks in the right-hand column?

1. Sanctuary   A. Never ____ before they hatch.

2. Meretricious   B. How much is that doggie ___?

3. Birmingham   C. ___ much.

4. Innuendo   D. ___ and a happy new year!

5. Conscience-   E. A ___ is worth two in the bush.
   stricken

RHYMING THE RHYMELESS

In the pages of *Phactum*, the newsletter of the Philadelphia Association for Critical Thinking, Elaine Brody composed the following clever clerihew:

William of Orange
Caught his fingers in a doorhinge.
As he pulled his finger loose,
He said "Hey! That's not orange juice!"

It has long been recognized, of course, that there are no common English words that rhyme with words like *orange*, *month*, *silver*, *bulb*, *wasp*, and *purple*. That, however, has not deterred word players like Ms. Brody from rhyming them. Here, for example, is Tom Lehrer's attempt at rhyming *orange*:

Eating an orange
While making love
Makes for bizarre enj-
Oyment thereof.

Willard R. Espy, in *Words to Rhyme With*, solved the problem this way:

The four eng-
Ineers
Wore orange
Brassieres.

No less than Stephen Sondheim worked *silver* into a poem as well:

To find a rhyme for silver
Or any "rhymeless" rhyme
Requires only will, ver-
Bosity and time.

And here are two impressive Victorian attempts to rhyme *month*. The second is by Christina Rossetti.

"You can't," says Tom to lisping Bill,
"Find any rhyme for month."
"A great mithtake," was Bill's reply;
"I'll find a rhyme at wunth."

How many weeks in a month?
Four, as the swift moon runn'th.

A more recent example by puzzle writer Francis Heaney goes like this:

If the last attempt to find a rhyme for "month"
Were numbered n,
It follows, then,
The one I'm writing now must be the n-plus-oneth.

The Persephone Is Ringing

Science fiction and mystery author Stephen Barr is responsible for these witty lines, based on classical words that are harder to rhyme than they appear.

The wicked shades
All go to Hades,
But not there gone
Is Hermione.
She sings, we hope,
With Calliope.
Under a proscenium
Resembling the Atheneum.

THE CHICKEN VARIATIONS

Everyone knows the old riddle about the chicken crossing the road. Here are four punny variations.

Why did the chicken cross the road halfway?

Why did the chicken cross the playground?

Why did the chicken cross the Möbius strip?

Why did the calf cross the road?

The Siege of Belgrade

In the section above on acrostics, I mentioned the abecedary psalms, which use the order of letters in the alphabet for their structure. There are many ingenious examples in English of abecedary verse, perhaps the most famous being Alaric Alexander Watts's "The Siege of Belgrade."

An Austrian army, awfully arrayed,
Boldly by battery besieged Belgrade.
Cossack commanders cannonading come,
Dealing destruction's devastating doom.
Every endeavor engineers essay,
For fame, for fortune fighting—furious fray!
Generals 'gainst generals grapple—gracious God!
How honors Heaven heroic hardihood!
Infuriate, indiscriminate in ill,
Kindred kill kinsmen, kinsmen kindred kill.
Labor low levels longest, loftiest lines;
Men march 'mid mounds, 'mid moles, 'mid
    murderous mines;
Now noxious, noisey numbers nothing, naught
Of outward obstacles, opposing ought;
Poor patriots, partly purchased, partly pressed,
Quite quaking, quickly "Quarter! Quarter!" quest.
Reason returns, religious right redounds,
Suwarrow stops such sanguinary sounds.
Truce to thee, Turkey! Triumph to thy train,
Unwise, unjust, unmerciful Ukraine!
Vanish vain victory! vanish, victory vain!
Why wish we warfare? Wherefore welcome were
Xerxes, Ximenes, Xanthus, Xavier?
Yield, yield, ye youths! ye yeomen, yield your yell!
Zeus', Zarpater's, Zoroaster's zeal,
Attracting all, arms against acts appeal!

The Five Airy Creatures

Jonathan Swift, who wrote *Gulliver's Travels*, also wrote this clever puzzle-poem:

> We are little airy creatures,
> All of different voice and features;
> One of us in glass is set,
> One of us you'll find in jet,
> T' other you may see in tin,
> And the fourth a box within.
> If the fifth you should pursue,
> It can never fly from you.

What are the five "airy creatures"?

Paradoxical Limericks

> There was a young girl in Japan
> Whose limericks never would scan.
> When someone asked why,
> She said with a sigh,
> "It's because I always attempt to get as many
>   words into the last line as I can."

> Another young poet in China
> Had a feeling for rhythm much fina.
> His limericks tend
> To come to an end
> Suddenly.

> There was a young lady of Crewe
> Whose limericks stopped at line two.

> There was a young man of Verdun.

The four preceding limericks, when I published them in my *Scientific American* column, prompted the British writer of comic verse J.A. Lindon to compose the following:

A most inept poet of Wendham
Wrote limericks (none would defend 'em).
"I get going," he said,
"Have ideas in my head.
Then find I just simply can't."

That things were not worse was a mercy!
You read bottom line first
Since he wrote all reversed—
He did every job arsy-versy.
A very odd poet was Percy!

Found it rather a job to impart 'em.
When asked at the time,
"Why is it? Don't they rhyme?"
Said the poet of Chartham, "Can't start 'em."

So quick a verse writer was Tuplett,
That his limerick turned out a couplet.

A three-lines-a-center was Purcett,
So when *he* penned a limerick (curse it!)
The blessed thing came out a tercet!

Absentminded, the late poet Moore,
Jaywalking, at work on line four,
Was killed by a truck.

So Clive scribbled only line five.

ONE-LINERS

Thanks to Dr. Michael W. Ecker for permission to reprint some choice aphorisms from his journal *Recreational and Educational Computing*.

A pessimist's blood type is always B negative.

Shotgun wedding: A case of wife or death.

Just because electricity comes from electrons does not mean that morality comes from morons!

Marriage is the mourning after the knot before.

A successful diet is the triumph of mind over platter.

A gossip is someone with a great sense of rumor.

They told me I was gullible ... *and I believed them*!

The mantra of every "yes-man": I came, I saw, I concurred!

Without geometry, life is pointless.

When two egotists meet, it's an I for an I.

It's not an optical illusion. It just looks like one.

I used to be indecisive, but now I'm not sure.

WHO WROTE SHAKESPEARE?

Robert Service wrote the following droll quatrain to support the conjecture that Francis Bacon was the true author of all of Shakespeare's plays:

> Said Jack McBrown to Tam McSmith,
> "Come on, ye'll pay a braw wee dramlet;
> Bacon's my bet—the proof herewith ...
> He called his greatest hero—*Hamlet*."

POOR JANET

There's more going on in this sad elegy that meets the eye.
Can you find the wordplay concealed in its structure?

Janet was quite ill one day.
Febrile trouble came her way.
Martyr-like, she lay in bed;
Aproned nurses softly sped.
Maybe, said the leech judicial
Junket would be beneficial.
Juleps, too, though freely tried,
Augured ill, for Janet died.
Sepulchre was sadly made.
Octaves pealed and prayers were said.
Novices with ma'y a tear
Decorated Janet's bier.

ENRICH YOUR WORD POWER

Try these linguistic catches on a patient friend.

I love the taste of *snew*.
What's *snew*?
Nothing much. What's new with you?

I love the taste of *updock*.
What's *updock*?
Who are you, Bugs Bunny?

I love the taste of a *henway*.
What's a *henway*?
About three pounds.

Mangled Prufrock

A whimsical challenge is to take a short lyric poem, or a passage from a longer poem, and rearrange the words (punctuation may differ) to make a different poem. Here, for example, is a passage from T.S. Eliot's "Love Song of J. Alfred Prufrock":

> Shall I part my hair behind? Do I dare to eat a
>   peach?
> I shall wear white flannel trousers, and walk
>   upon the beach.
> I have heard the mermaids singing, each to each.

And here is my rather crude rearrangement of the same passage:

> Upon the flannel beach I heard
> A singing part to "Hair."
> My mermaid's trousers have behind
> The white each peach shall wear.
> Each shall I walk to eat, and I ...
> I do dare!

Kubla Khan

The first five lines of one of Samuel Taylor Coleridge's most famous poems are a remarkable example of alliteration. The final two words of each line (disallowing the tiny *to* in line four) begin with the same letter.

> In Xanadu did Kubla Khan
> A stately pleasure dome decree;
> Where Alph, the sacred river, ran
> Through caverns measureless to man
> Down to a sunless sea.

Four All Who Reed and Right

This ode to problematic English plurals is well over a century old, yet is still popularly passed along in many variants.

We'll begin with a *box*, and the plural is *boxes*,
But the plural of ox should be *oxen*, not *oxes*.
One fowl is a *goose*, but two are called *geese*,
Yet the plural of *moose* should never be *meese*.
You may find a lone *mouse* or a nest full of *mice*,
Yet the plural of *house* is *houses*, not *hice*.

If the plural of *man* is always called *men*,
Why shouldn't the plural of *pan* be called *pen*?
If I spoke of my *foot* and show you my *feet*,
And I give you a *boot*, would a pair be called
    beet?
If one is a *tooth* and a whole set are *teeth*,
Why shouldn't the plural of *booth* be called
    beeth?

Then one may be *that*, and three would be
    those,
Yet *hat* in the plural would never be *hose*,
And the plural of *cat* is just *cats* and not *cose*.
We speak of a *brother* and also of *brethren*,
But though we say *mother*, we never say
    methren.

Then the masculine pronouns are *he*, *his*, and
    him,
But imagine the feminine *she*, *shis* and *shim*.
So the English, I think you all will agree,
Is the queerest old language you ever did see.

A Tense Time With Verbs

Richard Lederer composed this similar ode to verb conjugation, first seen in his 1989 book *Crazy English*, and reprinted many times since then:

> The verbs in English are a fright.
> How can we learn to read and write?
> Today we speak, but first we spoke;
> Some faucets leak, but never loke.
> Today we write, but first we wrote;
> We bite our tongues, but never bote.
>
> Each day I teach, for years I taught,
> And preachers preach, but never praught.
> This tale I tell; this tale I told;
> I smell the flowers, but never smold.
>
> If knights still slay, as once they slew,
> Then do we play, as once we plew?
> If I still do as I once did,
> Then do cows moo, as they once mid?
>
> I love to win, and games I've won;
> I seldom sin, and never son.
> I hate to lose, and games I lost;
> I didn't choose, and never chost.
>
> I love to sing, and songs I sang;
> I fling a ball, but never flang.
> I strike that ball, that ball I struck;
> This poem I like, but never luck.

I take a break, a break I took;
I bake a cake, but never book.
I eat that cake, that cake I ate;
I beat an egg, but never bate.

I often swim, as I once swam;
I skim some milk, but never skam.
I fly a kite that I once flew;
I tie a knot, but never tew.

I see the truth, the truth I saw;
I flee from falsehood, never flaw.
I stand for truth, as I once stood;
I land a fish, but never lood.

About these verbs I sit and think.
These verbs don't fit. They seem to wink
At me, who sat for years and thought
Of verbs that never fat or wought.

MORE "BETCHAS"

"I'll bet I can make you say blue." If the victim accepts the bet, ask, "What color is the sky?" When he answers "Red" (or some other color) say, "I win! I made you say *red*!" If he falls for the catch he will say, "But that wasn't the bet. You said you'd make me say *blue*." Of course you now have won.

Borrow a quarter, put it in your fist, and say, "I claim that by sleight of hand I've changed this quarter to a half dollar. Can I have the quarter if I'm wrong?" Without thinking just what your statement means, she may say, "Okay." If so, say, "I'm wrong," open your first, and pocket the quarter.

Wordsworth's Rainbow

One of Wordsworth's most famous poems is "The Rainbow":

> My heart leaps up when I behold
> > A rainbow in the sky:
> So was it when my life began;
> So is it now I am a man;
> So be it when I shall grow old,
> > Or let me die!
> The Child is father of the Man;
> And I could wish my days to be
> Bound each to each by natural piety.

Jeremy Morse noted that the poem is remarkable for its lexical brevity—the average length of each word in the poem is only 3.08 letters! He wrote the following version of the poem that contains *no* words longer than three letters:

Minimal Rainbow

> A joy it is for me to see
> > A bow set in the sky:
> So was it for me as a boy;
> So is it now I am a man;
> So let it be in my old age,
> > Or let me die!
> For man, I see, is son to boy;
> And I'd add day to day and so
> Go on and on as I was set to go.

David Morice couldn't resist producing a "translation" of the poem with much longer words—in fact, an average word length of almost eleven letters.

Maximal Rainbow

Ecstatically, happiness visualization
    Rainbow arranging skylights:
Existentially sprouting youthfully;
Currently experiencing adulthood;
Permitting sameness agelessly,
    Alternatively: Euthanasia's lifelessness.
Adulthood becoming offspring;
Additionally diurnally recreating
Continuance, preparationally departing.

TERSE VERSE

Ogden Nash's "Further Reflections on Parsley" notes, in its entirety, that

Parsley
Is gharsley.

He also wrote this immortal "Geographical Reflection":

The Bronx?
No thonx.

Nash's most famous short poem is probably his 1930 observation that

Candy
Is dandy
But liquor
Is quicker.

But it grew three words longer in 1968 when Nash added this postscript: "Pot is not."

The following oft-published couplet makes Nash seem positively wordy by comparison:

Fleas

Adam
Had 'em.

George Plimpton says that, at a Harvard commencement lecture, Muhammad Ali was asked to compose a poem on the spot. His attempt was even shorter:

Me:
Whee!

Plimpton was also responsible for publishing, in a 1969 anthology, the following poem by Aram Saroyan (the son of novelist William). It was this single word in the center of a blank page:

lighght

Saroyan received $500 from a National Endowment for the Arts grant for the publication of the poem. This seemed to some an exorbitant payment for writing a single (misspelled) word, and the resulting kerfuffle was the start of the congressional criticism of the NEA that continues to this day.

This poem is less controversial, but for sheer brevity, it's hard to beat. Eli Siegel published it in the *New York Evening Post*'s *Literary Review* in 1925, and used to recite it to great acclaim at the Village Vanguard, the famed Manhattan jazz club.

One Question

I.
Why?

When William Cole wrote an essay on "One-line Poems and Longer, but Not Much" *(New York Times Book Review,* December 2, 1973), the review later published (January 13, 1974) a letter from G. Howard Poteet in which he proposed one-letter poems: "Thus, my work includes the most evocative of all poetic letters, 'O.' Further, there is the egocentric poem ('I'), the poem of pleasure ('M'), the scatological verse ('P'), and the somnambulistic bit of poesy adapted from the comics, ('Z')."

I suggest we take this a step further with the following poem titled "Simplicity":

.

No one can say my poem does not have a point. Of course we can write an even simpler poem, completely pointless, with the title "Ultimate Simplicity." It goes like this:

Answers to Chapter 3

Seven Puzzle Poems

1. This "square" poem reads the same forward as it does when the first word of each line is read in order, followed by the second word of each line, etc.

2. The poem (with byline) is in the shape of its subject, a chair.

3. The initials of the first 25 words (not counting the words in parentheses) are the letters of the alphabet in order—minus the letter *U*. According to the sign quoted at the beginning of the poem, the bus holds 26 people. Updike is describing only his *fellow* passengers, so the poet himself is the missing *U*.

4. The first two letters of each line, read in sequence, spell "peculiar acrostic."

5. The last letter of each word, including the words in the title, is the first letter of the following word.

6. The only vowel in this poem is *i*.

7. This is a word palindrome; it reads the same forward as it does when the words are read in reverse order.

More Ridiculous Riddles

1. With some root beer, two scoops of ice cream, and a hippopotamus.

2. A cherry that works nights as a grape.

3. Harry Elephante.

4. A chair, a bed, and a toothbrush.

5. A baseball team with the measles.

6. Nothing.

7. Someone who wonders if there is a dog.

8. In their sleevies.

9. They taste funny.

10. Tep on the brakes, toopid!

ACROSTICS

In the Edgar Allan Poe acrostic sonnet, "An Enigma," the lady's name is Sarah Anna Lewis. This can be read by taking the first letter of the first line, the second letter of the second line, and so on through all fourteen lines. Mrs. Lewis was a Baltimore poet whose efforts were considered "rubbish" by Poe but to whom he was indebted for financial aid. The sonnet is not as well known as his other acrostic, "A Valentine," in which another lady's name is similarly concealed. The word "tuckermanities" in line 10 of "An Enigma" refers to the conventional, sentimental work of Henry Theodore Tuckerman, a New York poet and author of the day.

In J.A. Lindon's Christmas poem "To Those Overseas," the first line can be read in two other ways: by reading the first word of each line, and the nth word of each nth line.

PUNCTUATED POEMS

Read the poem like this:

> There was an old lady and she
> Was deaf as a P-O-S-T.
> A boy and girl behind a tree
> Were K-I-S-S-I-N-G.

A Loony Lexicon

1. C ("thank you very")
2. D ("merry Christmas")
3. E ("bird in the hand")
4. B ("in the window")
5. A ("count your chickens")

The Chicken Variations

1. To lay it on the line.
2. To get to the other slide.
3. To get to the same side. (Or, if you trust your comic delivery, you may prefer to give the answer as "To get to the other ... hey, wait a minute.")
4. To get to the udder side.

The Five Airy Creatures

The little "creatures" are the five vowels: *a, e, i, o, u.* Jonathan Swift called them "airy" because they are actually made of air—the air that comes out of your throat and makes the vowel sounds.

Poor Janet

The twelve lines begin with the abbreviations for the months of the year: *Jan, Feb, Mar,* and so on.

4
Brief Brain Busters

1. There is a common four-letter word that ends with the letters *eny*—and it's not *eeny*. What word is it?

2. Name a common eight-letter word that begins and ends with the letters *he*.

3. Now name a common word that contains the consecutive letters *adac*.

4. Put three letters in front of *ergro* and the same three letters at the end of *ergro* to make a word. What is the word?

5. Write 11030 on a piece of paper, then change it to the name of a type of person by adding two straight line segments.

6. Three common, unrelated words begin with *dw*. Can you name them?

7. Is it true that day begins with *d* and ends with *e*?

8. A vacationer said to his fellow guests at breakfast one morning, "10SNE1?" Decode what he asked them.

9. Print the letters W, W, A, U, and L on five individual squares of paper, as above. Can you arrange them in a row to spell the name of an animal?

10. Spell out the names of the first 999 numbers, from *one* to *nine hundred ninety-nine*. What's the most common letter of the alphabet you'll never use?

11. An accountant noticed that there were two adjacent double letters in the word *balloon*, and then wondered if there was a common word with *three* adjacent pairs of double letters. There is. What is it?

12. Draw a tiny line somewhere on the letter sequence A B C D E to make a familiar five-letter word.

13. Rearrange the letters of *ocean* to spell a different watery word.

14. BUT = TRUTH denotes a phrase in a poem by a famous British poet. Can you quote the phrase?

15. What is the next letter in this sequence: WITNLIT?

16. "Remember," a dentist said to a patient, "that you must *rush*, *loss*, and *wish* every day." What must you do to each italicized word so that his advice makes sense?

17. Who might say, "IMNNDNNATP"?

STILL MORE LINGUISTIC CATCHES

Ask someone to say "milk" three times. When he or she finishes, ask, "What do cows drink?" Most people will say "milk," but of course adult cows don't drink milk. They drink water.

Ask a friend to think of any word and you will use your psychic powers to write *the exact word* on a sheet of paper. After you're told the word, show that you have written down "the exact word."

"How many of each kind of animal did Moses take on the ark?" Whether the answer given is "Two" or anything else, say, "Wrong! It was Noah who built the ark."

18. In number theory, Christmas (Dec. 25) and Halloween (Oct. 31) are closely related. How?

19. *October* and *Sunday* have no letters in common. Can you find a second month/weekday pair that has the same property?

20. Find an anagram for *chesty*.

21. NBS is the acronym for the National Bureau of Standards, the old name for the agency that officially keeps America's time. Appropriately enough, how is this acronym related to the months of March and October?

22. And what month has a similar relationship to the words *hip* and *tub*?

23. I claim there are only two *R*'s in Robert Richardson's name. Am I right?

24. What U.S. state has the last four letters of its name the same as the first four letters of its capital?

25. What word becomes shorter when you add two letters to it?

26. Jeremiah Jason, a radio disc-jockey, has a business card that reads:

J. JASON, D.J., FM-AM.

What is unusual about that sequence of letters?

27. Name ten words that identify things that go on the feet, each with a name that begins with *s*.

28. Name eight words starting with *sn* that relate to the nose.

29. I once met a lady who claimed she came from Mars. Was she telling the truth?

30. How can you make a Greek cross with five matches?

31. Name a U.S. state that ends in *g*, and a state that ends in *h*.

32. What word contains within it the consecutive triplet *spb*?

33. Can you think of two words of opposite meaning that, if placed after the word *SLOW* on a highway sign, will make a two-word sign that gives the same instructions to motorists?

34. *Continuum* has two adjacent *u*'s. Can you name two other common words with the same property? What is the shortest common English word to have three *u*'s that are *not* consecutive? (It's much less unusual than scientific terms like "Cuculus" or "cumulus.")

35. The two rebuses below represent meteorological forecasts:

BABS
O

WETHER

Can you decipher them?

36. There are only two common words (and one uncommon word) that are each an anagram for one of the seven colors of the rainbow. What are they?

37. Cross out seven letters in FIVE PLUS SIX PLUS SEVEN so as to leave a statement that gives eighteen, the correct sum.

38. A barber named Robinson said he would rather cut the hair of two men named Smith than cut the hair of one man named Jones. Why?

39. Rearrange the letters of the phrase *new door* to spell one word.

40. *Tennessee* starts with *ten*. Name a state that starts with 10 and another state that ends with 10.

41. What word do all Harvard graduates spell incorrectly, and what word do they always pronounce wrong?

42. Which one of the following words doesn't belong on the list: uncle, cousin, mother, sister, father, aunt?

43. Charles Lutwidge Dodgson, better known as Lewis Carroll, considered using the pen name "Edgar Cuthwellis," and later published three poems under the byline "R.W.G." Can you think of why he chose those pseudonyms?

44. Carroll once challenged a friend to scramble ABCDEFGI—the first nine letters of the alphabet, *H* excepted—to spell a hyphenated word. Can you do it?

45. In what common English word is the letter *f* pronounced like *v*?

CLASS CLOWNS

"Two negatives," explained the logic professor, "always make a positive, but two positives never make a negative." Sarcastic voice from the rear of the class: "Yeah, right."

Teacher: Where was the Declaration of Independence signed?
Student: At the bottom.

Teacher: Our bodies have a nose to smell with and feet to run with.
Student: I must be made all wrong, teacher. It's my nose that runs and my feet that smell.

Teacher: Name two pronouns.
Student: Who, me?
Teacher: Correct.

Philosophy professor: Explain the difference between a stoic and a cynic.
Student: A stoic is a big bird and a cynic is where we wash dishes.

46. Mary's mother has three daughters. The oldest daughter's first name is April, and the middle daughter's first name is May. What's the youngest daughter's name?

47. What's heavy when taken forward, and not when taken backward?

48. Utah has four letters,
    Up has but two.
    Can you spell them
    Without using *U*?

49. What common substance might I decide to call HIJKLMNO?

50. What do these words have in common: worth, mouth, fast, and zest?

51. Remove a letter from *zebra* and rearrange the other letters to make the name of another beast.

52. The letters *rstu* appear consecutively in the alphabet. Is there an English word that contains them together in the same order?

53. How quickly can you find a word that contains the letter sequence *achach*? What about *tantan*?

54. A large American city and a large Canadian city share the following property: the first two letters of their names are the same as the last two letters. Name the cities.

55. Can you name a world city whose name, when written out in mixed case, has three dots in a row? What about a country name with the same property?

56. What large U.S. city has the name of a mammal?

57. What nation has the name of a bird?

58. What letter is not in the name of any U.S. state?

59. If that one's too easy: what's the only state capital that does not share a letter with its state?

60. What eight-letter word contains the consecutive letters *abc*, in that order? If you get stuck on that, try thinking of the name of a jazz legend with the same property.

61. What woman's first name is suggested by this slanting line: / ?

62. This sequence has ten letters: TAHUREDENE. Cross out three to leave the name of a famous British poet.

63. What animal's name is twice as long in the plural form as in the singular?

64. Which day of the week has an anagram that's a familiar word?

65. *Sugar* is a common word in which the first letter *s* is pronounced as "sh." Can you think of another?

66. What letter comes after AB in the alphabet? The answer isn't C!

67. What familiar English word might alternately be spelled E10100010001000UNI100ATXN?

68. What word, commonly seen on coins and bills, becomes its own opposite when its middle two letters are reversed?

69. What part of the body has a 10-letter name in which each of its five letters appears exactly twice?

70. What's the only English word that can be produced by joining two three-letter abbreviations for months of the year (as in *janfeb*)?

71. Rearrange the letters of *elation* to spell a part of a person's body.

72. What three-letter word for an animal becomes a two-letter word for a different animal if you cross out its initial letter?

73. What five-letter word for a large animal becomes a small animal if you alter a letter?

74. Change one letter in the word *shuffle* to make a name for something you can eat.

75. In the two blank spaces of *a* ___ *d* ___ *t* put the same three-letter word to make a nine-letter word.

ANSWERS TO CHAPTER 4

1. Deny.
2. Headache.
3. Headache (again).
4. Underground.
5. Hobo.
6. Dwarf, dwell, dwindle.
7. Yes, *day* starts with *d* and *ends* starts with *e*.
8. "Tennis, anyone?"
9. Turn one *W* upside down to make an *M*, rotate the other one to make an *E*, and rotate the *U* to make a *C*. The word is CAMEL.
10. The letter *a*.
11. Bookkeeper. If the accountant's assistant were a subbookkeeper, it would increase the number of doublets to four.
12. Make the *C* into an *O* to spell ABODE.
13. Canoe.
14. "Beauty (B-U-T) is truth," from Keats's "Ode on a Grecian Urn."
15. *S*. The letters are the first letters in the statement of the problem.
16. Add a new first letter to each word, to spell *brush, floss,* and *swish*. (Thanks to Raymond Young.)
17. An Indian in a tepee.

18. The number 25 in "Dec." (decimal) notation is the same as 31 in "Oct." (octal) notation.

19. June, Friday.

20. Scythe.

21. Shift each letter of NBS back one step in the alphabet to get MAR, and forward one step to get OCT.

22. Shift HIP forward six steps in the alphabet to get NOV. Shift TUB backward six steps to get NOV.

23. Yes, Robert Richardson's name has only two uppercase *R*'s.

24. Vermont, Montpelier. (Thanks to Iris Craddock.)

25. Short.

26. The letters on the card are the initial letters of the twelve months of the year, starting with June.

27. Among the possible answers: shoes, socks, slippers, sneakers, sandals, snowshoes, skis, stockings, skates, spats, slingbacks, and stilettos.

28. Among the possible answers: snout, snort, sniff, sniffle, snivel, snore, sneeze, snooty, snoop, snuff, snuffle, and snot.

29. Yes, Mars is a town in Pennsylvania.

30. Stick four matches in the Greek person's ear and light them with the fifth.

31. Wyoming, Utah.

32. Raspberry.

33. Slow up, slow down.

34. Vacuum and muumuu, unusual.

35. The first rebus is "two degrees above zero." (The two degrees are the BA, bachelor of arts, and BS, bachelor of science.) The second is "a bad spell of weather."

36. Lube and genre (and onager).

37. Eliminate letters to leave IV PLUS IX PLUS V. (4 + 9 + 5, in Roman numerals.)

38. He would make twice as much money.

39. *New door* anagrams to *one word*.

40. Iowa, Ohio.

41. Incorrectly, wrong.

42. *Cousin* is the only word without a gender.

43. Edgar Cuthwellis is an anagram for "Charles Lutwidge." R, W, and G are the fourth letters of the three names Charles Lutwidge Dodgson.

44. Big-faced.

45. Of.

46. Mary, of course.

47. The word *ton*.

48. Yes, it's easy to spell *them* without using the letter *U*.

49. Water, because HIJKLMNO is "H to O."

50. Replace their first letters to spell *north*, *south*, *east*, and *west*.

51. Bear.

52. Understudy (less common answers like *overstuffed* are also possible).

53. Stomachache, instantaneous.

54. Miami, Toronto. (Thanks to Iris Craddock.)

55. Beijing, Fiji.

56. Buffalo.

57. Turkey.

58. Q.

59. Pierre, South Dakota.

60. Dabchick, Cab Calloway.

61. Eileen (*I* lean).

62. Cross out *THREE* to leave *AUDEN*.

63. Ox(en).

64. Monday (dynamo).

65. Sure.

66. The letter *e* comes after *ab* in *the alphabet*.

67. EXCOMMUNICATION, substituting letters for Roman numerals and O for zero.

68. United, untied.

69. Intestines.

70. Decoct.

71. Toenail.

72. Fox.

73. Moose and mouse.

74. Souffle.

75. Amendment. (Thanks to Will Shortz for this and the preceding four teasers.)

5

ENTANGLED WORDS AND NUMBERS

ALPHAMAGIC SQUARES

A magic square is an $n \times n$ array of numbers arranged such that the numbers in each row, column, and diagonal sum to the same total.

Surely the most fantastic 3×3 magic square ever discovered is one constructed by Lee Sallows, a British electronics engineer who works for the University of Nijmegen in the Netherlands (seen at right).

5	22	18
28	15	2
12	8	25

It would be hard to guess its amazing property, so I'll tell you what it is. For each cell, count the number of letters in the English word for its number, then place these counting numbers in the corresponding cell of another 3×3 matrix. For example, *five* has four letters, so 4 goes into the top left corner of the new matrix. Here is the result:

4	9	8
11	7	3
6	5	10

Not only is it another magic square, but its entries are the consecutive integers from three to eleven! Sallows calls the first square the *li shu* (punning on the famous *lo shu* magic square of Chinese legend as well as on his own first name), and the second square its alphamagic partner. His computer investigations of alphamagic squares in more than 20 languages are reported in his two-part article

"Alphamagic Squares" in *Abacus* (Vol.4, 1986, pp. 28–45, and 1987, pp. 20–29, 43).

Mamikon Mnatsakanian, an Armenian physicist, noticed a further coincidence. The rows, columns, and diagonals in the first square sum to 45, and those in the second square to 21. Both those numbers are spelled with nine letters, mirroring the nine cells in each square.

EVEN MORE CLASSIC CLASSROOM CATCHES

While trading knock-knock jokes, say to a friend "I just thought of a great knock-knock joke. Start me out." Without thinking, your friend may say, "Knock knock." If so, you reply, "Who's there?" It will catch the friend speechless!

Tell someone that you know of a device that enables one to see right through the walls of a house. When he expresses surprise and asks for details, say, "It's called a *window*."

Tell someone you can speak every foreign language except Greek. Add: "Name any language and I'll prove it." Regardless of the language you are given, say, "Sorry, but that's all Greek to me."

CRYPTONUMBER

The first 10 letters of the alphabet, ABCDEFGHIJ, can represent, in cryptogram fashion, the spelled-out name of a number. Note that each encrypted letter stands for a different plaintext letter, so the answer is a number whose name contains 10 letters, each occurring exactly once. (I've removed any spaces or hyphens.) What's the number?

The 46th Psalm and Shakespeare

It has long been noted that the 46th word of the 46th Psalm is *shake*, and the 46th word from the end (ignoring the final *Selah*) is *spear*! Moreover, Shakespeare was 46 years old in 1611, the same year that the King James translation was published in England. There's more! The 14th word is *will*, and the 32nd word from the end (counting the final *Selah* this time) is *I*, followed by *am*. The sum of 14 and 32 is 46.

Some have seized upon these oddities to claim that Shakespeare himself may have assisted in the work on the King James translation. Alas, these are all amazing coincidences. J. Karl Franson, writing in *Word Ways* (August 1994), revealed that an earlier English translation of the Bible, by Richard Taverner, already included all the numerology given above! The Taverner translation was in 1539, some 25 years before Shakespeare was born. The coincidences are more astonishing when one learns that the wording of Psalm 46 is not identical in the two translations!

The Whistling Agent

A passenger asked a ticket agent how long a train would be at the platform, and the agent replied, "2222222." "What, are you the train's whistle?" the passenger indignantly replied. What did the agent's answer mean?

College Mixer

At a college mixer the students were asked to pin their names on the front of their clothing. A girl pinned 317537 on her blouse. Another girl pinned 31573. What were their names?

If you can solve that riddle, you'll know what number their friend Ollie Lee requested on his license plate when he bought a new car.

A Consecutive Coincidence

The last letters of the integers eigh*t*, nin*e*, and te*n* spell *ten*.

Honest Numbers

Write down any number. Count its letters, then write the word for that number. Count the letters in the new word, and again write down the name of that number. Continue in this way, counting each number word to get a new number. After a short number of steps you will always reach the same endpoint. It is what mathematician Michael Ecker named a black hole because its four letters end the sequence. *Four* is the only number with this property, namely that of counting the letters in its own name.

Because of that fact, I once described the number 4 as the only "honest number" in English. Readers suggested many ingenious ways to produce more honest numbers in English. "How about two cubed?" asked Morris H. Woskow, or "twelve plus one," "twenty minus five," "minus four squared," and so on. Walter Erbach sent a list of 40 examples that included "one-half of thirty," "square root of nine hundred sixty-one," and "integral of $x\, dx$ from sixteen to eighteen." Michael Burke and Norman Buchignani suggested such exotic phrases as "the largest prime less than thirty" and "the odd integer between thirty-eight and forty." Paul C. Hoell came up with a 62-letter phrase, "cube root of two hundred thirty-eight thousand three hundred twenty-eight," but this was topped by Robert B. Pitkin's 101-letter phrase, "Seventh root of one zero seven trillion two thirteen billion five thirty-five million two ten thousand seven hundred one."

Two correspondents, W.M. Woods and Malcolm R. Billings, each showed that such number names are infinite in number. The phrase "added to ten" has ten letters, therefore we can write an infinity of honest names merely by repeating this phrase as many times as desired, e.g., "four added to ten, added to ten, added to ten …"

Such contortions are unnecessary to find honest numbers in other languages. In his books *Language on Vacation* and *Beyond Language*, Dmitri Borgmann finds international examples from Italian (*tre*), Spanish (*cinco*), and a dozen other languages. It's possible, he found, to list consecutive honest numbers from *o* (*one* in Middle English) to *kuusteistkummend* (*sixteen* in Estonian).

CRYPTARITHM

A popular form of verbal math is the cryptarithm (a portmanteau word coined by combining the words "cryptogram" and "arithmetic"). A cryptarithm is a simple mathematical equation in which each digit, 0 through 9, has been substituted for a different letter of the alphabet.

For example, my friend Harry Hazard of Princeton, New Jersey, discovered during the days of the Johnson administration that you could write the then-president's name as a multiplication problem:

$$
\begin{array}{r}
\mathsf{L\ Y\ N\ D\ O\ N} \\
\times \qquad\qquad\qquad \mathsf{B} \\
\hline
\mathsf{J\ O\ H\ N\ S\ O\ N}
\end{array}
$$

There is a unique substitution of digits for letters that gives the problem a correct numerical form. Can you find it?

When I first posed this problem in *Scientific American*, two independent readers, Douglas G. Russell and Edward C. Devereux, discovered that there is also a unique solution to a similar multiplication problem (which can be seen on the following page).

$$.\text{M A R T I N}$$
$$\times \underline{\hspace{4cm}} .\text{A}$$
$$.\text{G A R D N E R}$$

The dots are decimal points. Not only does this have a unique solution, but no other middle initial can be substituted for *A.* to give a pattern with a solution. It would be pleasant to report that my middle initial is *A.;* unfortunately I have no middle name.

GROANERS

1. What are "gegs"?

2. What's the opposite of "not in"?

3. What kind of nut makes you want to sneeze?

4. What's the difference between a bus driver and a bad cold?

5. The alphabet goes from A to Z. What goes from Z to A?

6. What can go up a chimney down, but can't go down a chimney up?

7. Why are dogs such bad dancers?

8. What did the mommy buffalo say to her son when he left for school?

9. What part of London is in France?

10. Who is bigger, Mr. Bigger or baby Bigger?

Dozens of clever sentences, even poems, have been constructed for remembering the digits of famous irrational numbers such as π, e, and the square root of 2. The number of letters in each word of the mnemonic gives the digits. For π, the best known mnemonic, constructed by British astronomer Sir James Jeans, is: "How I want a drink, alcoholic of course, after the heavy chapters involving quantum mechanics."

When the teacher said "π r squared" (the formula for the area of a circle given its radius r), Tommy raised his hand. "No, teacher. Pie are round. Cake are square!"

In spite of Tommy's remark, here are some ways to show that pies are indeed square. Using the code $A = 1$, $B = 2$, $C = 3$, and so on, the letters of *PIES* add to 49, a square number. *PIE À LA MODE* adds to 81, another square. *RAISIN PIE* sums to 100 (10^2), and both *COCONUT PIE* and *ESKIMO PIES* total 121 (11^2). The *P* of *PI* is 16 (4^2), while *I* has a value of 9 (3^2). The sum of 16 and 9 is 25 (5^2), and the product of the two numbers (144) is another square ... as is the product of any two squares, admittedly. Dividing 9 by 16 gives .5625, and 5625 is another square (75^2).

I once perpetrated the following verse:

> Pi goes on and on and on,
> And e is just as cursed.
> I wonder, "How does pi begin,
> When its digits are reversed?"

Write the word *pie* like this:

ⱶＩƐ

Hold the page up to a mirror and you'll see the first three digits of pi!

My friend Mamikon Mnatsakanian carried this curiosity further. Write pi to seven decimals:

31415926

In the mirror you'll see the phrase "As ezi [as] pie."

> "Which transcendental number do you like best,
>   pi or e?"
> "I prefer pi," she replied palindromically.

Shown below are the capital letters of the alphabet. Cross out all those with left-right symmetry—that is, letters that look the same in a mirror. (In our diagram, we've grayed them out instead.) The remaining letters form groups whose number of letters, taken clockwise, gives 31416.

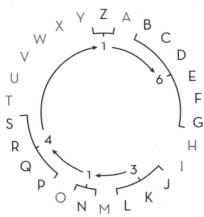

If you divide the circumference of a spherical squash by its diameter, what's the result? Pumpkin pi. (Thanks to John Evans.)

In Book II, Chapter 9 of H.G. Wells's novel *The War of the Worlds* a sentence begins "For a time I stood regarding...." By an amazing coincidence, the number of letters in each word gives pi to six digits! This was discovered (how he discovered it beats me!) by Michael Keith of Richmond, Virginia. He includes it in an unpublished work titled *How I Wish I Could Recollect Pi*—a title whose word lengths, you may have already noticed, give pi to seven digits.

MATHEMADICS

Pictures that represent words in some puzzling way are called *rebuses*. Mathematicians who enjoy wordplay have designed their own cousin of the rebus, in which mathematical terms are designed to illustrate their own meaning. I coined the term "mathemadics" to refer to these pictograms. The examples below should give you the basic idea. Can you design others?

TOPOLOGY
———
BIS   ECT

GRAPH
———

S I N E W A V E
———

DILATATION
———

LIMIT POINT
———

E^x PONENT
———

SUB script
———

TRANSLATION
———
PE RI OD IC

A
D
D
———

MULTIPL
Y
———

R O T A T I O N
———

S
E
C
INTER
I
O
N
———

(MAT RIX)

An ABC Curiosity

One thousand, as noted in Chapter 4, is the smallest number with a name that contains the letter *a*. Can you find the smallest number that contains *b*? What about *c*? And what letter appears in no number name between 0 and 99, but appears in the name of *every* number between 100 and 999,999?

An AEIOU Curiosity

What's the smallest number in which all six vowels, including *y*, appear? (Remember that *and* is not a proper part of any number name.) And how high must you count to get to the smallest integer whose name includes the five vowels in *aeiou* order?

Final Number

The alphabetical list of the English spellings for the integers 0 through 1,000 begins like this: eight, eight hundred, eight hundred eight, eight hundred eighteen, eight hundred eighty, and so on. The last entry, of course, is zero. Can you name the 1,000th, or next-to-last, number on the list?

Last of the Romans

You can pose a similar question about the Roman numerals. The digits themselves are now the letters. The sequence begins: C, CC, CCC, CCCI, CCCII, and so on. The Romans had no zero, and so the last number is the 1,000th one. Can you determine what it is?

Going Forward

The only number name with its letters in reverse alphabetical order is *one*. What's the only number name whose letters are in *forward* alphabetical order?

A Linguistic Proof

Here's a novel way to prove that 11 + 2 – 1 = 12. Write down the word *eleven*. Add *two* to make *eleventwo*. Now subtract (cross out) the three letters of *one*. This leaves *elevtw*, an anagram of *twelve*!

THREE RIDDLES

Jim and his sister Joan discovered a jar of cookies in the kitchen cupboard and had a snack. Their score: Jim 81, Joan 812. How many cookies did they eat in total?

What U.S. coin doubles in value if you take away half?

Why is 6 afraid of 7?

Odd vs. Even

Odd is spelled with an odd number of letters.
Even is spelled with an even number of letters.
Using the cipher $a = 1$, $b = 2$, etc., *odd* sums to 23, an odd number. *Even* sums to 46, an even number that is twice 23.
The cipher sum of *even plus even* is 160, an even number. The sum of *odd plus odd* is 114, an even number. And the sum of *odd plus even* is 137, an odd number. (Thanks to Owen O'Shea for the above curios.)
All prime number are odd except two. This makes two, paradoxically, a very odd prime.

A Dicey Coincidence

The faces of a die add to twenty-one pips. Using the cipher $a = 1$, $b = 2$, $c = 3$, etc., the four letters of *dice* also add to 21. (Thanks again to Owen O'Shea.)

A Strange Number

Arrange the ten digits in alphabetical order: 8549176320. This artificial number has peculiar properties. Divide it by 5 and you get 1709835264, a number with all ten digits. Divide it once more by 5. Result: 341967052.8. Again, all ten digits!

Now divide by 4. The quotient is 85491763.2, the original alphabetical sequence with no zero at the end. Amazing? No, because 5 × 5 × 4 = 100, dividing by which moves a decimal point two spaces to the left.

Divide 8549176320 by 2718, the first four digits of *e*, and you get a number that begins with 314, the first three digits of pi!

Drop a line in care of the publisher if you can discover any other curious properties of 8549176320.

NON SEQUITURS

To err is human; to moo, bovine.

Fact: the speed of time is one second per second.

Protons have mass? I didn't even know they were Catholic.

Seventy percent of American adults believe statistical claims even when no source is cited.

When you've seen one shopping center you've seen a mall.

There are only 10 kinds of people: those who understand binary numbers, and those who don't.

Time flies like an arrow, but fruit flies like a banana.

Calendar Cubes

In a magazine column, I once asked how to put digits on the twelve faces of two cubes so that they could be placed side by side to give any day of the month. (See my *Mathematical Circus*, page 186.) Now imagine a similar set of three cubes with lower-case letters on each face. How would you distribute letters on three cubes so that they can be turned and rearranged to spell, in lower case, the first three letters of any month?

Equations in Verse

Why is this a limerick?

$$((12 + 144 + 20 + 3\sqrt{4}) \div 7) + (5 \times 11) = 9^2 + 0$$

Because, as textbook author Jon Saxton has pointed out:

> A dozen, a gross, and a score,
> Plus three times the square root of four,
> Divided by seven,
> Plus five times eleven,
> Is nine squared and not a bit more.

A simpler number limerick was composed by palindromist Leigh Mercer:

$$1,264,853,971.2758463$$

> One thousand two hundred and sixty-
> Four million eight hundred and fifty-
> Three thousand nine hun-
> Dred and seventy-one
> Point two seven five eight four six three.

GOOD ADVICE

Don't sweat petty things—or pet sweaty things.

Two can live as cheaply as one … but only for half as long.

Reading while sunbathing can make you, well, red.

Persons stupid enough to believe in phrenology should go have their heads examined.

Dream in color: it can be a pigment of your imagination.

Puns are bad, but poetry is verse.

Taking Attendance

Ten students enrolled for a class in number theory. To aid in remembering their names, the professor seated them in the following order. Can you determine the basis for his ordering?

Don Edwards
Robert Woods
Edith Reed
Rolf Oursler
Jeff Ives
Jessi Xander
Rose Ventnor
Leigh Thompson
Toni Nesbit
Pete Norris

ANSWERS TO CHAPTER 5

CRYPTONUMBER

The number is *eighty-four*.

THE WHISTLING AGENT

Knowing that the train was scheduled to arrive at the station at 1:58 p.m. and stay for twenty-four minutes, he replied, "Two to 2:00 to 2:22."

COLLEGE MIXER

The two nametags said *ELSIE* and *LESLIE*. The girls had just pinned them on upside down! Ollie Lee's license plate number is 33731770, which also spells his name when turned upside down.

CRYPTARITHM

LYNDON × B = JOHNSON has the unique answer of 570140 × 6 = 3420840, and .MARTIN × .A = .GARDNER has the unique answer of .124867 × .2 = .0249734.

Groaners

1. Scrambled eggs
2. "In." You didn't say "not out," did you?
3. Cashew!
4. The bus driver knows his stops, and the cold stops his nose.
5. Zebra
6. An umbrella
7. Because they have two left feet
8. "Bison!"
9. The letter *n*
10. The baby, because he's a little Bigger

An ABC Curiosity

B first appears in *one billion*, and *c* in *one octillion*. The letter that appears in every number from 100 to 999,999 is *d*.

An AEIOU Curiosity

The first number with all six vowels is 1,025. You won't get to an *aeiou* number until 1,084.

Final Number

Two hundred two is the next-to-last number in the list alphabetically.

Last of the Romans

The last entry is XXXVIII, or 38.

Going Forward

Forty is the only number name in alphabetical order.

Three Riddles

1. Only two: Jim 81 ("ate one"), and Joan 812 ("ate one too").
2. A half dollar becomes a dollar when you take away *half*.
3. Because 789 ("seven ate nine").

Calendar Cubes

The "unfolded" cubes shown below demonstrate how three cubes can be given lower-case letters so that the cubes can be arranged in a row to spell the first three letters of any month. Note that this is only possible because *u* and *n* and *p* and *d* are inverses of each other.

Taking Attendance

The numbers *one* through *ten* are concealed between their first and last names. For example, *one* is concealed in the italicized letters in D*on E*dwards.

6
CONJURING WITH WORDS

PRECOGNITION

Alice in Wonderland begins with the following words:

> Alice was beginning to get very tired of sitting
> by her sister on the bank, and of having nothing
> to do: once or twice she had peeped into the
> book her sister was reading ...

Select any word before the colon. Spell the chosen word by counting forward, one word at a time, and note the word on which the spelling ends. Again spell *that* word, counting forward until you reach a new word. Continue spelling words in this fashion until you can't go farther in the excerpt.

Using my powers of precognition, I predict the final word is *sister*!

The same trick works with Julia Ward Howe's famous "Battle Hymn of the Republic," which begins:

> Mine eyes have seen the glory of the coming of
>   the Lord;
> He is trampling out the vintage where the
>   grapes of wrath are stored;
> He hath loosed the fateful lightning of his
>   terrible swift sword;
> His truth is marching on.

Follow the same procedure as before. Select any word in the first line, from *Mine* through *Lord*. Believe it or not, no matter what word you picked to start the chain, the final word will be *marching*!

MIRROR MAGIC

Timothy and Rebecca were at a party where each guest had his or her name, printed vertically, pinned to their clothing.

```
T                    R
I                    E
M                    B
O                    E
T                    C
H                    C
Y                    A
```

When they stood in front of a mirror they noticed a strange thing. The mirror reversed REBECCA but had no effect on TIMOTHY! Hold this page up to a mirror and you'll see this is true.

CHOICE	PURPLE	COOKBOOK	WATER
WAR	DIED	TIGER	ECHO
ICEBOX	SQUARE	BOO-HOO	TABLE
TURTLE	HIDE	LARGE	DECIDED
OBOE	ROSE	CHOKED	PIG

Half the words in the above chart are on black cells, half on white. Turn this page *upside down* and hold it in front of a mirror. Amazingly, the mirror reverses all the black words but leaves the white words unchanged! You should have no trouble figuring out why this works.

Only one of the following sentences is false. All the others are true:

1. CARSON WAS BORN CHRISTMAS EVE 1809— LIVED TO THE AGE OF 58
2. BUFFALO BILL WAS BORN IN 1846—HIS BIRTHPLACE WAS SCOTT COUNTY, IOWA
3. HICKOK DIED DEC 3 1883—DOC BEECH DECIDED HE CHOKED
4. CUSTER WAS KILLED AT LITTLE BIG HORN MONTANA IN JUNE 1876
5. CROCKETT OF TENNESSEE MET DEATH AT THE ALAMO IN THE YEAR 1836

A mirror will instantly identify the false sentence. Just hold this page upside down in front of a mirror and look at the reflection. The false sentence will be the only one you can read! (Thanks to Frank Brady.)

HOW COME?

1. A man has no grand children, yet he has several great grandchildren. How come?

2. Mudville won a baseball game 12 to 3, yet not a single man on the team touched third base. (And yes, it was a men's team.) How come?

3. A cowboy rode into town on Friday, stayed three days, and left on Friday. How is this possible?

4. There are three people in a room. They are not men, not women and not children. So who are they?

5. A newspaper in Omaha reported that a man had married seven women, all still living and none divorced. Yet the man was never arrested for bigamy. How come?

Fold Two Words

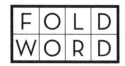

Crease a rectangle of paper four times as shown above. In the squares write the words *FOLD* and *WORD*. Now fold the sheet along the creases. Do this any way you like until you have formed a square packet of eight leaves. With scissors, trim the edges of the packet to make eight separate squares.

Spread the squares on the table. Four will have their letter sides uppermost. Arrange the four letters to form a word.

Put these squares aside, then turn over the four face-down squares. Make another word with them. You may be surprised to discover that no matter how you folded the rectangle, the two words you spell will be *FOLD* and *WORD*!

Fold and Trim

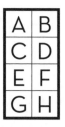

Fold a sheet of paper so that when you open it flat the creases will form eight rectangles. Write in these boxes the first eight letters of the alphabet in the order shown above.

Now fold the sheet along the creases in any manner you like to make a packet with the eight rectangles together, like the leaves of a book. With scissors, trim away all four edges of the packet. This will leave you with eight separate paper rectangles.

Spread the eight pieces on the table. Four letters will be faceup and four facedown. Try to form a word with the face-up letters. If you can't make a word, turn the pieces over and try again with the other letters.

I'll now predict the word you will form. It's *HEAD*. (Thanks to Max Maven.)

Impossible Animals

Ask someone to write down the following five words:

1. The name of a nation that begins with *D*.
2. An animal whose name starts with the second letter of the nation.
3. The animal's color.
4. An animal whose name begins with the *last* letter of the nation.
5. A fruit that begins with the last letter of the animal named in the previous step.

Tell your friend to concentrate hard on visualizing the five answers. Then say, "Aha, I have it! But wait a minute—there *are* no gray elephants or orange kangaroos in Denmark!"

The outcome isn't guaranteed, but the odds are good they'll have come up with the five answers you predicted.

Three Words

If you're successful with the previous prediction, try this similar one. Ask the person to write down the following words:

1. A wild beast.
2. The largest city of a foreign country.
3. A vegetable.

You can't be sure of your prediction that the words will be *lion*, *Paris*, and *carrot*, but you'll be surprised by how often those are right.

A Bird in the Hand

Here's a third prediction that seems to work more often than not. Ask someone to write down names for the following:

1. A U.S. state that starts with A.
2. An insect that begins with the last letter of the state.
3. A wild animal that begins with the last letter of the insect.
4. A bird that begins with the last letter of the animal.

Dramatically announce that "The bird is a robin!" (If you're lucky, the four answers will have been *Alaska* (or *Alabama* or *Arizona*—just so long as it's not *Arkansas*), *ant*, *tiger*, and *robin*.

A Smart Trick

Write down a three-digit number that does not contain zero and in which the first and last digits differ by more than one. Reverse the digits to make a new number. Subtract the smaller number from the larger. Reverse the digits of the difference, and add the reversed number to the previous one.

For example, suppose you start with 387. Reverse to get 783. 783 − 387 = 396. The number 396 reverses to make 693, which is then added to 396.

Translate the final sum to letters by using the following chart:

S	M	A	R	T
1	2	3	4	5
6	7	8	9	0

You can predict the word that results. It will always be *STAR*.

Counting Cards

From a full deck, deal cards one at a time to the table, spelling A-C-E, one card for each letter. Continue dealing to spell T-W-O, T-H-R-E-E, and so on until you spell K-I-N-G. Surprise! The spelling terminates on the last card of the deck!

Mysterious Order

Arrange the thirteen spades (or the cards of any other suit) from ace to king in the following face-down order with the queen on top:

$$Q-4-A-8-K-2-7-5-10-J-3-6-9$$

Spell A-C-E by moving a card from the top of the packet to the bottom with each letter, except for the last card which you turn faceup and deal to the table. It will be the ace. Repeat by spelling T-W-O, placing the two on the table. Continue spelling *three, four, five,* and so on, each time transferring a card from top to bottom with every letter you spell, except for the last card of each word which is dealt face up on the table. In this way you spell each card from ace to king!

You need not recall the proper sequence of cards at the start. Just follow the procedure in reverse! Start with the king. Put the queen on top as you start the spelling with Q, then transfer cards from *bottom* to *top* as you continue with U-E-E-N. After you've spelled all thirteen cards, from king through ace, the packet will be in proper order for spelling ace to king.

Spooky

Think of any number in the twenties; that is, any number from 20 through 29. Call the number *x*. Deal *x* cards to the table. Pick up the packet and note its bottom card. Let's say it is the queen of hearts. Add the two digits of x to get the number y. Deal y cards from the top of the packet back to the top of the deck and place the rest of the packet on top.

Dealing from the top of the deck, spell *THIS IS A SPOOKY TRICK* by dealing a card for each letter. Turn over the last card dealt. It will be the queen of hearts!

ANNOY YOUR FRIENDS!

What's the sum of 5Q and 5Q?
10Q.
You're welcome.

What do you call that thing you're wearing on your left foot?
A shoe.
Gesundheit!

Shout, "It's running down my back!" When asked, "What is?" calmly answer, "My spine."

SPELL YOUR NAME

Form a pile by dealing to the table any number of cards from one through ten. Alongside it form a second pile by dealing the same number of cards. Look at the top card of the remainder of the deck and replace it. Say it is the king of hearts.

Choose one of your two piles on the table and put it on top of the king. Next deal to the table a card for each letter in your name provided your name has more than ten letters. (If your name is too short, include your middle name, or use the name of someone else, such as the name of a person watching you do the trick.) After spelling the name, put the tabled packet on top of the deck. Pick up the remaining pile and put it also on the deck.

Now spell your name again by dealing the cards to the table. The last dealt card will be the king of hearts!

The trick always works in spite of the fact that each of the two original piles had an arbitrary number of cards, and regardless of the length of the name used for the final spelling.

THE PYRAMID SPELL

Place cards ace through six faceup on the table like this:

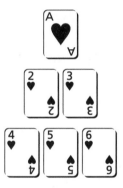

Pick up the cards along the three diagonals as follows. Put the face-up 6 on the 3, then both on the ace. Place the three cards on 5, then on 2. Finally, the packet goes on 4. Flip over the whole stack; from top down, the cards will be ordered 4-2-5-A-3-6.

Hold the packet face down and spell A-C-E by moving a card from top to bottom for each letter. Turn over the *next* card and place it aside. It will be the ace. Continue by spelling *two, three, four, five,* and *six,* always turning the *next* card face up and putting it aside. The spelling will produce all six cards in sequence, an amazing coincidence. Thanks to Colm Mulcahy, a mathematician at Spelman College, for discovering this surprising word trick.

Last Two Cards Match

One of the most mysterious of all spelling tricks, not well known today even to magicians, was invented by mentalist Howard Adams. He published it in his 1984 book *OICUFESP*, which was rare for many years but is now back in print. (Can you figure out the meaning of its title?) I urge you to get a deck of cards and astonish yourself by following these instructions.

Remove from the deck any five cards and their mates. A mate is a card of the same value and color. For example, the queen of hearts is the mate of the queen of diamonds. Call the five cards ABCDE and their mates abcde. Arrange the cards in the order A-B-C-D-E-a-b-c-d-e. Place the packet on a table and cut it one or more times. Hold the packet of ten cards facedown and deal five cards to form a pile, thereby reversing their order. Put the remaining five cards facedown to form a pile alongside the cards just dealt.

Now you are going to spell the words in the phrase *Last two cards match*. Pick up *either* pile and spell the letter *L* by moving one card from the top of the pile to the bottom. Replace the pile on the table alongside the other one. Again, randomly select one of the two piles. Pick it up and move a card from top to bottom to spell *A*. Replace the pile, then pick up either pile to spell *S*. Do the same for *T*.

You now have two piles face down on the table. Remove the top cards of each pile. Without showing their faces, put the face-down pair to a vacant spot on the table, one card overlapping the other.

Repeat the random procedure for selecting a pile and spelling *T*, then *W* and *O*. After spelling *TWO*, again remove the top cards of the two piles, and place them aside, facedown, near the pair previously put aside.

In the same way randomly select piles for spelling *cards* and *match*. After each spelling of a word, put the top cards of the two piles to one side, facedown by the other pairs.

Two face-down cards remain on the table. Turn them over. Surprise! In spite of all the random choices, the cards match! The prediction, "Last two cards match," has been fulfilled!

Is the trick over? Not by any means. There is a second climax even more astounding. Turn over all the pairs you placed aside. Each pair consists of matching cards!

The trick rests on subtle combinatorial principles. Of course other words can be substituted for the four in the prediction, provided they are spelled with four-three-five-five letters, or those numbers with five, four, three, and two added to them to make words with nine-seven-eight-seven letters. Do you see why?

Answers to Chapter 6

How Come?

1. The man's children aren't grand, but all his grandchildren are great. (Thanks to Ray Smullyan)

2. There's not a "single man" on the team because all the players are married.

3. Friday was the name of the cowboy's horse.

4. A man, a woman, and a child

5. The man was a minister, and he married seven women to their husbands.

Last Two Cards Match

The letters *OICUFESP*, when read aloud, sound like "Oh, I see you have ESP."

Word lengths of nine-seven-eight-seven will work just as well because you just add an additional cycle through the card packets, bringing the same cards back to the top. (The packets shrink in size from five to two as the trick proceeds.)